Lawnmowers
& Grasscutters
A Complete Guide

Brian Radam

The Crowood Press

First published in 2011 by
The Crowood Press Ltd
Ramsbury, Marlborough
Wiltshire SN8 2HR

www.crowood.com

British Library Cataloguing-in-Publication Data
A catalogue record for this book is available from the British Library.

ISBN 978 1 84797 252 1

Acknowledgements
The author and publisher gratefully acknowledge the help of the following in putting
together this book: Su Radam, Ian Britstone, Lawnmower Warehouse Southport, Atco,
Qualcast, Bosch, Husqvarna, Flymo, Hayter, AL-KO, Ransome, Briggs & Stratton, Tecumseh,
Robomow, Aspen Fuels, NGK Spark Plugs, Champion Spark Plugs, Mike Wareing,
Madeleine Fyne, Martin Garner, Malcolm Pye, Bill Addison, Andy Hall, Joanne Hall, Vicky
Capstick, Nora Pilkington, John Barnard, Keith Wright, Graham Allan, Derek Martlew,
Patrick Paynter, Rainbow Cards Ltd.

Disclaimer
The author and the publisher do not accept any responsibility in any manner whatsoever
for any error or omission, or any loss, damage, injury, adverse outcome, or liability of any
kind incurred as a result of the use of any of the information contained in this book, or
reliance upon it.

Typeset by Jean Cussons Typesetting, Diss, Norfolk
Printed and bound in India by Replika Press Pvt Ltd

Contents

PREFACE

Over the years the art of lawnmower manufacturing has remained something of a mystery and lawnmowers have on occasions baffled the most professional of engineers. These machines, once destined only for royalty and the rich gentry, can now be acquired by everyone. Through time there have been many thousands of the best design engineers, the best computer technology and many registered patents bestowed on mowers – some look good enough to launch a mission to Mars – but amazingly, for the most prestigious lawns, golf greens, football and cricket grounds throughout the world today, without exception, the preferred choice is still the first lawnmower design, invented in 1830.

Although I was deprived as a child, having only a back yard, my lawnmower journeys have led me up the garden path of life and taken me through many gates. When opened, each gate has brought me to encounter people from all over the world, each one with their own lawnmower story, from the most passionate enthusiasts to the most entertaining comedians like Ross Noble, who devoted one of his best two-hour shows to talking non-stop to a 500-strong audience about the humorous side of the lawnmower, ending with the words, 'If I die now I will be happy.'

I have talked to great-grandsons of a scytheman who cut grass before the lawnmower had been invented. They told me that many scythemen didn't get a scythe until they had stopped growing, as the scythes were made to measure. They could achieve a bowling green finish with a scythe, a skill we have lost nowadays as we have lawnmowers to do the work. If they wanted the grass to be a quarter of an inch longer they would put a leather sole under their shoe to raise their body. Their party trick would entail picking a cigarette paper up with the blade without cutting any grass.

These extremely skilled men often cut in gangs of six or eight, followed by women and children collecting all the cuttings. They worked unsociable hours, either very early in the morning or late at night, as the best time to cut grass with a scythe is when the grass is moist with dew. A back-breaking job, and they knew that within three or four days they would have to do the same thing all over again.

Once a lovely old lady from the Lake District in her ninety-seventh year was chatting away and she said, 'I remember when I was six years old and our coachman who drove me around in the horse and carriage used the same horse to pull the lawnmower around the lawns, and these are the leather boots he put on the horses' feet so as not to mark the ground.' She showed me the boots: they were obviously donkeys' years old, but nevertheless in excellent condition and superbly made. She went on, 'Our coachman didn't want the horse to poop on the newly mown lawn so he guided it over to a convenient bush where he did a little whistle; on hearing this the horse would relieve itself and fertilize the plants at the same time.' I could not help noticing embossed right on the front of the shoes a large second-quality stamp.

It fascinated me how much better the quality must have been in those days, an era when you could have first- or second-quality leather boots for your donkey or pony to tramp around a muddy field.

There have been many other memorable highlights: meeting a Lancashire brewer making a unique beer called 'Half-Cut', and experimenting with a professor making a grass-flavoured beer (in a UK survey, freshly mown grass was voted as the nation's favourite smell). Talking about the grass roots of lawnmower racing with Sir Stirling Moss and the fascination of antique lawnmowers with Michael Aspel. John Sergeant arriving at the Lawnmower Museum on a garden tractor. Receiving Paul O'Grady's lawnmower decked with pink furry handles and a leopard-skin chassis – if Lily Savage were to have a lawnmower it would definitely be that one. Laughing whilst filming in the BBC Blue Peter Garden demonstrating the first robot mower with Konnie Huq, where the sheep dog thought it was a sheep and kept rounding it up each time it moved! A chance encounter on a deserted beach in the Caribbean, where a sculptor was selling his carved fish and birds. When asked if he could create a lawnmower there was a slight hesitation; then with confidence he replied, 'Yeah mon, no problem! – But what's a lawnmower?' He went on to make a superb lawnmower, carved from a single piece of driftwood, complete with handle engine and blade. If Fred Flintstone had a mower it would have been this one! It is now a unique exhibit and part of the British Lawnmower Museum collection.

The Museum has had some wonderful donations: Brian May from the famous rock band Queen rang up one day and said, 'Brian May here, would you like my old lawnmower?' The answer was, without hesitation, 'Yes, please!' His machine sits alongside those of other celebrities such as HRH Prince Charles and Princess Diana, Paul O'Grady, Hilda Ogden, Alan Titchmarsh, Nicolas Parsons, Roger McGough, Richard Bacon, Joe Pasquale and many more who have donated all sorts of gardening artefacts.

My interest in lawnmowers led me to find out about the terms 'Shanks's Pony' (a term meaning 'travelling on foot', derived from the name of an early horse-drawn lawnmower) and the cricket term 'in the slips' – this relates to when the grass on the outfield was kept short by sheep, causing the players to slip in their deposits!

Then there is the bizarre story Albert Pierrepoint's lawnmower: Britain's most famous executioner, he hung 400 people, for which he was paid fifteen pounds per hanging. The lawnmower he bought cost fifteen pounds, and I wondered who had paid for his lawnmower? It is now exhibited in the British Lawnmower Museum, where it hangs from the ceiling; strange but true.

Whilst making the film *Lawnmowerworld* with the daunting task of choosing suitable music, we received an unexpected phone call: 'You don't know me, but I'm a direct descendant of the inventor of the lawnmower and I've written and sung a song.' We were amazed to hear that the lyrics of the 'Lawnmower Song' told the actual history of the lawnmower (we were thinking of using 'The Green Green Grass of Home' by Tom Jones or 'One Man Went to Mow').

Once when chatting about the weather to a reverend gentleman arriving from Australia and visiting Great Britain on a cloudy day, he said it was like 'coming from the garden into the garage,' and aptly I thought, 'if this book can keep more people in the garden, spending less time in the garage, it's done it's job.' A world without grass and lawns would be a much greyer place!

ABOUT THE AUTHOR

Ex-racing champion Brian Radam, often known as 'the Lawn Ranger', was born in Lancashire above Southport's first DIY shop, started by his father in 1945. Amongst some of the things his father did were repairs of locks, safes and lawnmowers. Although not even having any garden as a young child, it was a treat for Brian to sit in a lawnmower grassbox and be pushed up and down the street by his father. By the age of ten he was regrinding lawnmower blades on a lathe and he became intrigued and fascinated with anything mechanical. He met his wife-to-be when they were ten years old: it was a lovely summer's day in the park, she was lying on the grass on her back in the sunshine, whilst she siphoned petrol out of a council mower. He knew she was the right one for him (in his dreams!).

He initially worked as an apprentice for The Atco Lawnmower Company repairing 425 lawnmowers every week, where the original engineers collected and delivered lawnmowers on Rudge motorbikes with sidecars.

He became a Master Locksmith and is a Fellow of the British Locksmiths Institute, practising the secret mechanical art of locks, keys and safes. Then in 1988 he officially opened the unique and now world-famous British Lawnmower Museum with over 200 restored lawnmowers (part of 800) that had been previously scrapped because they were too old, too expensive to repair and with no parts available, and would have been more sensibly disposed of in a crusher. Fortunately he saved these special historical machines destined for Lawnmower Heaven and the rare contraptions were restored for all to see. He has travelled from the north to the south of England giving unique talks called 'One Man Went to Mow' about the lighter side of the lawnmower's history, a time when British engineering was the best in the world, and including forgotten stories that would not have been passed on without talking to people who lived and worked cutting grass over a hundred years ago.

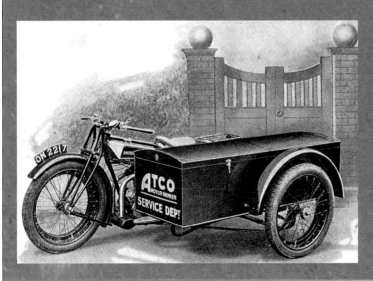

Atco service engineers travelled all over the UK on Rudge motorbikes with sidecars to carry the lawnmowers and all the tools with which to service them.

INTRODUCTION

On a recent visit to the British Lawnmower Museum, Robbie Coltrane observed, 'People spend more time cutting grass than making love.' Although this seems incredible, a recent survey found that on average a man walks 220 miles and a woman walks 48 miles over a lifetime simply cutting grass! Lawnmowers are a very British obsession but surprisingly one that is rarely admitted to. Unfortunately it's just not the sexiest subject. When Stephen Fry was asked to present a programme on lawnmowers he said, 'Well, I don't know anything about lawnmowers!' This led me down to my local pub to ask the question, 'Who's got a lawnmower?' Do you know, not one person admitted to owning one, although between us we buy a million machines every year.

British Lawnmower Museum postcard. The lawnmower shown is a Budding.

The aim of this book is to give an interesting guide for advice, hints and tips, and to help save money, time and energy on mower care and maintenance. The guide explains why lawnmowers give problems and how to eliminate them. It helps you to choose the right machine for your lawn, to prolong the life of most common domestic mowers, and perhaps take out the frustration when the mower won't start on your only sunny day off. This guide covers many popular domestic lawnmowers and grasscutters. The content is not too technical and you will not require a degree in engineering to follow it. The lawnmower is the most neglected of all household products with neighbours compounding the problem by borrowing it when theirs has just broken. Eighty per cent of the time it will break whilst cutting the grass and most likely when the grass is long and needs cutting. Even the most house-proud of us often has a neglected mower. Unlike other mechanical machines, some modern domestic mowers have been made with serious economical restraint and have been designed to the lowest cost.

The important safety instructions in this book cover normal conditions. Unusual, unforeseeable circumstances may occur; in these situations common sense, caution and care are needed by the person operating and maintaining the equipment. Every care has been taken that the information in this book is correct. However, environmental issues, design, and equipment specification are constantly changing during manufacture and differences may be found. In view of this, neither the author nor publisher can accept any liability for any loss, damage or injury as a result of change, omission or error in the book. The advice given is the quickest, easiest and least expensive way to keep or get your mower back up and running and keep it going. The guide may not replace a full restoration service but gives explanations and reasons why and how things are made and work and should give a better understanding when your mower breaks down.

All the descriptions in this book refer to the left, the right, the front, the rear or back as viewed from the normal operating position. The term 'lawnmower' in most cases refers to cylinder machines; the term 'grasscutter' usually refers to rotary machines.

With all types of lawn and grasscutting machinery – especially where spare parts and imported models are involved – there are continual changes in manufacture, so please bear in mind that there's always an 'exception'!

CHAPTER ONE

A SHORT HISTORY OF THE LAWNMOWER

Some people say the most beautiful word in the English language is 'lawnmower'; some say the opposite. The population of Great Britain can be divided into two groups: those who love cutting the grass, and those who think it's a pain in the gr-ass. We buy well over a million mowers every year, but whichever side of the garden fence you're sitting on the grass will still keep growing, especially if it's above 11°C (52°F). This book should make those who find mowing the lawn a nuisance a little more comfortable with the experience, and it will help to make it as enjoyable and relaxing as it was originally intended to be. As its inventor proclaimed in 1830 for the very first time, 'Gentlemen may find my machine an amusing and healthy exercise.'

The first person known to have had a lawn was a Roman called Pliny the Younger, in 62 ad. But the word 'lawn' didn't appear in the English lan-guage for another 500 years until the mid-1500s. The word 'lawn' means 'open clearing in woodland'. A more modern definition would be closely cut grass, especially around a house. Grass itself is a family of long, narrow plants with jointed stems, which can be grown as a lawn or pasture. Some people ask what the difference is between grass and turf. Turf is the surface layer of grass and its roots, often referred to as a horse race track. Whilst all lawns are made of grass, not all grass can be classified as a lawn!

Officially Edwin Beard Budding of Stroud in Gloucester (also less well known for inventing a gun and an adjustable spanner) invented the lawnmower in 1830. He didn't originally set out to make the lawnmower; it came about while he was working in a textile mill. The mill owner had received an order for guardsman's uniforms and wanted the cloth to be perfect, so he asked Edwin Budding to make him a machine to cut all the tufts and bobbly bits off the nap of the cloth. What he invented was a simple cylindrical-cut-

ABOVE: **Pliny the Younger had the first recorded lawn in 62 AD.**

RIGHT: **A country gentleman with the first lawnmower, invented by Edwin Beard Budding in 1830.**

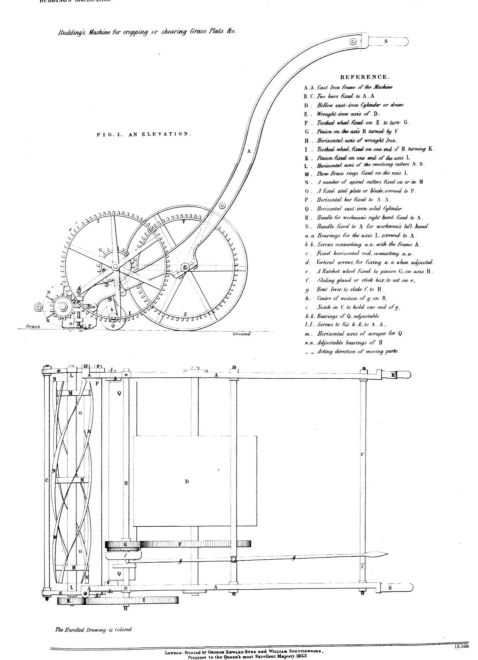

A.D. 1830, AUGUST 31. № 6081.
BUDDING'S SPECIFICATION.

(1 Sheet)

Budding's Machine for cropping or shearing Grass Plats &c.

FIG. 1. AN ELEVATION.

REFERENCE.

A.A. Cast Iron frame of the Machine
B.C. Two bars fixed to A.A
D . Hollow cast-iron Cylinder or drum
E . Wrought-iron axis of D .
F . Toothed wheel fixed on E to turn G .
G . Pinion on the axis B turned by F
H . Horizontal axis of wrought Iron.
I . Toothed wheel fixed on one end of H turning K .
K . Pinion fixed on one end of the axis L.
L . Horizontal axis of the revolving cutters N.N .
M . Three Brass rings fixed on the axis L.
N . A number of spiral cutters fixed on or in M
O . A fixed steel plate or blade screwed to P.
P . Horizontal bar fixed to A.A.
Q . Horizontal cast-iron solid Cylinder.
R . Handle for workman's right hand fixed to A.
S . Handle fixed to A for workman's left hand
a.a. Bearings for the axis L. screwed to A.
b.b. Screws connecting a.a. with the frame A.
c . Fixed horizontal rod, connecting a.a.
d . Vertical screws, for fixing a.a. when adjusted.
e . A Ratchet-wheel fixed to pinion G, on axis H.
f . Sliding gland or click box to act on e,
g . Bent lever, to slide f, to H .
h . Centre of motion of g, on B.
i . Notch in C. to hold one end of g,
k.k. Bearings of Q, adjustable
l.l. Screws to fix k.k, to A.A.
m . Horizontal axis of scraper for Q
n.n. Adjustable bearings of H
←. Acting direction of moving parts.

Grass

Ground

The Enrolled Drawing is Colored.

LONDON: Printed by GEORGE EDWARD EYRE and WILLIAM SPOTTISWOODE,
Printers to the Queen's most Excellent Majesty. 1853.

Edwin Budding's original patent of 1830, one of over 600 patents and blueprints from 1799 held at the Lawnmower Museum. Original patents were written on parchment type paper. (Courtesy of the archive department at The British Lawnmower Museum, Southport.)

ABOVE: Ransome's 'New Automaton' *c.* 1867.

RIGHT: Edwin Budding's all cast-iron lawnmower, 1830. Budding's words about his invention were: 'Gentlemen will find my machine an amusing and healthy exercise'.

SHANKS'S PONY

ABOVE: The author with a 24in Greens of Leeds and London horse-drawn mower.

Many early lawnmowers were classed in man sizes, starting with a 'one-man mower', then a 'man-and-boy mower', and a 'two-man mower'. Next were the donkey mower, pony mower and horse-powered lawnmower, up to an elephant mower. The most famous of these animal-powered machines was called 'Shank's Pony', from Arbroath in Scotland, and this is where the term 'Shanks's Pony' comes from. The horses and ponies were shod with leather boots so as not to mark the lawn. Examples of these early machines made from cast iron and wood with iron gears can be seen in The British Lawnmower Museum in Southport, Lancashire, Great Britain.

LEFT: Work animals were shod with leather boots (in four sizes: donkey, pony, horse and shire horse) so as not to mark the lawn.

Green's *Silens Messor* ('silent cutter'). Green's entered the Guinness Book of Records for keeping this same model for seventy years.

ting mechanism, consisting of a revolving cylinder blade over a stationary fixed blade. Edwin Budding found it also cut the grass very efficiently so he went to the engineer John Farabee next door and together they started making his strange contraption.

Although we take the lawnmower for granted nowadays it wasn't plain sailing for Budding. People thought him a madman and lunatic for inventing such a contraption (who would possibly want one?), so he had to test the machine at night in order that no one would see him.

The principle Edwin Budding invented has not changed to this day and is still the main cutting mechanism for formal lawns all over the world. He made 600 lawnmowers in six years; then a large agricultural company called Ransome bought him out. They called some of their new lawnmowers 'The New Automaton'.

Green's of Leeds and London also manufactured many models; some of the early models from the mid-1800s had Latin names such as *Multum in Parvo*, meaning 'much with little'. Another was Green's *Silens Messor* meaning 'silent cutter'. It was the first time the new invention, the 'block chain' had been used, making the machines much quieter and lighter than previous models, which used cast-iron gears. The new

invention consisted of a link made up of a solid block of steel followed by an open link into which the sprocket tooth engaged. The effect of this made the teeth on the early sprockets much wider apart compared to a modern chain. The main benefits of the block chain over cast iron gears were quieter operation and lighter weight. This model is in the Guinness Book of Records for remaining unchanged for seventy years. Made using superb British engineering – the best in the world at the time – the lawnmower was perfect, so why change it?

Just before the internal combustion engine was invented, Sumner built the steam-powered lawnmower in 1893 at the Leyland Works in Lancashire (British Leyland was more famous for manufacturing the Mini). It was a breakthrough in technology and a labour-saving machine, requiring only one man instead of two or more. It cost just 1d an hour to run and, unlike a horse, you did not need to feed it. Three models were made, starting with the lightest 25in model weighing 9 cwt (457kg) and costing £60. A 30in

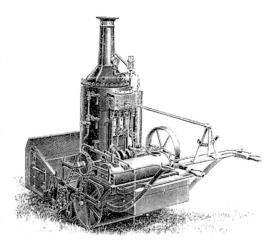

An 1890s steam lawnmower weighing nearly a ton. This machine was a breakthrough in technology (built before the internal combustion engine) with the added advantage of one-man operation.

The first ride-on lawnmower by Ransome, purchased in 1904 for the Cadbury Chocolate workers' sports field. Pictured in the background is Bourneville House, which is still there today. For the first time a horse was not needed to pull the lawnmower. Ransome say this is where the term 'horse power' came from, and the machine was certainly 'a cut above the rest' at the time.

model weighing 14cwt (711kg) cost £75, and the monster 36in model weighing 17cwt (864kg) cost £90. These weights did not include the huge weight of water for the boiler and the large capacity of grass it could hold. The cost of these machines would have been a small fortune in the 1890s, considering the average wage in England at that time was less than £2 per week.

The first petrol ride-on lawnmower was made by Ransomes in 1904 and was bought by Cadbury Chocolates and Bourneville House, for their workers' sports field. Ransomes say it's where the term 'horse power' originated.

In the meantime in America in 1907, an electrical engineering graduate by the name of Stephen Briggs was designing a gasoline engine, not knowing that it was destined to become the most popular small mower engine throughout the world; together with Harold Stratton, he produced the Briggs & Stratton small combustion engine and founded the Outboard Marine Corporation.

There were several prestige motorbike and gun companies making lawnmowers, such as Royal Enfield, whose machines had superb gears and a motorbike-style kick start. They were advertised with the phrase 'Built like a Gun', and some of them were used to mow the velvet courts of the world-famous Wimbledon Tennis Club.

Other famous manufacturers included Vincent, who produced one of the world's best motorbikes famous for holding the land speed record;

RIGHT: **A very rare example, this Vincent mower features a fuel tank integrated with the handle.**

BELOW: **Royal Enfield Lawnmower logo, 'Built like a Gun', showing their famous cannon.**

the lawnmowers featured a unique handle, which also doubled as the fuel tank and toolbox. Velocette motorbikes made a superb twin-cylinder engine mower. Hawker Siddeley, more famous for making aeroplanes, produced mowers, as did Jerram and Pearson Precision Engineering from Leicester. These, manufactured in aluminium (rather than cast iron), were probably the highest quality and most expensive lawnmowers in the world: British engineering at its best. Advertised as 'the Rolls Royce of lawnmow-

ers' during the mid-1920s, the company also made the castings and parts for Rolls Royce. Eventually Rolls Royce had a big influence in the company but unfortunately they closed down in the 1970s. Dennis, more famous for making fire engines, manufactured superb lawnmowers from the 1920s onwards; they also bought the castings and machine tools from Jerram and Pearson and continue to use their superb designs today. There are many other lawnmower manufactures too numerous to mention – these are just a few.

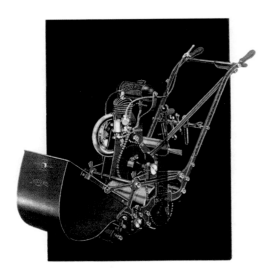

ABOVE: **The Atco Standard (in a rare black-painted advertisement). This was the first mass-produced petrol motor lawnmower (1921). Many featured a propeller for cooling the engine.**

ABOVE: **Atco Standard from the mid-1920s, complete with promotional advertising ladies.**

RIGHT: **The 12in Atco Standard cost 19 guineas. Features included the new open cast-iron chassis, starting crank handle, and tool rack. Engineers on motorbikes and sidecars backed up a nationwide service network.**

Atco – who still make lawnmowers today – were originally called 'The Atlas Chain Company', as they previously made chains for ships. Their famous lawnmower came about when the horse that pulled the lawnmower around the factory grounds died. The managers got together and asked, 'is there not another way of cutting the grass without getting another poor horse?' They designed and built the first mass-produced petrol lawnmower in 1921 at a breakthrough price of 19 guineas. It featured a unique cast-iron open frame, a crank handle start, built-in toolbox and a propeller to cool the engine. Atco set up a nationwide network with service depots all over the UK with trained service engineers calling on customers on motorbikes with sidecars. Atco made many different types of products, and around the Second World War they manufactured outboard motors for boats (some were advertised passing the Statue of Liberty – a similar one was used in the James Bond film *Moonraker*), the carriage for the famous 'bouncing bomb' and also a car, now a very rare piece of British motoring history fetching many thousands of pounds, and highly sought after by avid enthusiasts. The car was taken from their 1939 lawnmower chassis and engine with the cutter and rollers removed; it was an ideal inexpensive vehicle with which to train people to drive, repair, service and maintain a car. Many people learned to drive in the Atco Car, some at Preston police HQ in Lancashire during the Second World War; it was this car that was the catalyst and spark for many budding British racing drivers including Le Mans and Grand Prix champions. Now there is an elite Atco Car Owners' Club for those people who have the privilege of owning one, keeping this small part of British motoring heritage alive.

Many famous manufacturers, including Atco, Dennis, Ransome and Royal Enfield, have mowed – and still mow – the hallowed football and cricket grounds and lawn tennis courts all over the globe. Many of these companies achieved a Royal War-

Brochure for the Atco two-stroke lightweight lawnmower, c. 1956. These models were advertised as consuming one pint of fuel per hour. The 12in model would cut 750 square yards in an hour; the 14in model 1,000 square yards; the 17in 1,250 square yards; and the 20in model would mow 1,500 square yards per hour.

rant to the King and Queen. The German company Bosch now owns the Atco, Qualcast, Suffolk and Webb lawnmower companies.

All these lawnmowers are not to be confused with the rotary grasscutter – a much more modern machine, which didn't become popular until the 1930s (when lawnmowing was still very much a 'man' thing). One of the first rotary machines was called the Autosythe from Shay of Basingstoke. The Autosythes had an extremely robust non-rust alloy chassis with large (9-inch diameter) but narrow wheels and slick rubber

A futuristic advertisement for the early Flymo, 1965.

tyres; other models had two strong, wide aluminium front wheels with a steel rear roller at the back. Shay made their own two-stroke engine with early models, fuelled by 'petroil' (petrol and oil mixture). Starting was achieved by wrapping a rope around a large, strong aluminium pulley fixed on top of the engine flywheel. The unique blade system had a large disc with three multi-position 1.5-inch circular blades fixed to the rim; these could be moved round six times to the next sharp radius when blunt. On the four-wheel models the petroil tank also cleverly doubled as the handle.

The most famous name of the rotary type is from an innovative company called Flymo, a household name now, and synonymous with grasscutting (people often say they have a Flymo but have another type of mower or brand, just as a vacuum cleaner is often called a Hoover). A gentleman called Karl Dahlman went to a Brussels inventors' fair and made a fortune with Sir Christopher Cockerell's hovercraft design: effectively an engine on a bin lid. But it wasn't plain sailing for The Hovercraft Lawnmower Company, as in 1965 Flymo had a similar problem to Edwin Budding's 135 years earlier: people wondered what the strange contraption was. It did not look like a lawnmower, and it was the first mower to be made out of plastic. Anything in plastic was deemed very flimsy in the 1960s, although in fact the Flymo chassis is virtually indestructible, being made of similar material to a police riot shield. They had a hard job selling the machines so they

tried a few strategies, one being to employ a number of representatives who knocked on doors, demonstrated the machine to the lady of the house and encouraged them to have a go with this new, light and easy-to-use mower. They also surveyed thousands of housewives and asked what colour the mower would be, given a choice. The overwhelming answer was orange, which is why they are still orange to this day.

The robot mower was introduced in 1995, a lifestyle product aimed at those who can't cut, won't cut, or just hate cutting grass – or who like to be in charge of cutting the grass but don't cut it. Who knows what will come in the future? Perhaps new solar-powered laser mowers with no moving parts will be guided by GPS satellite to the areas that need cutting, and reduce the grass to tiny particles and moisture before recycling it back to the earth.

LAWNMOWER TRIVIA

- The fastest mower in the world was driven on 23 May 2010 for the Guinness Book of Records land speed record and reached a speed of 87.833mph. It was driven by Donald Wales, grandson of Sir Malcolm Campbell, who in 1924 held the world land speed record of 146mph.
- The mower that covered the most miles in twelve hours on a bumpy grass track travelled over 300 miles, at the British Lawnmower Racing Association annual twelve-hour race.
- The record for cutting the most grass in the shortest time is one acre per minute.
- Mikhail Kalashnikov, famous for inventing the AK47 machine gun, always said he was happy he'd invented the gun but always wished he'd invented something useful like the lawnmower.

CHAPTER TWO

CHOOSING THE RIGHT MACHINE

There are many thousands of different mowers on the market and many different methods of cutting grass, from a sheep or scythe, through cylinder, rotary, hover, or reciprocating knife machines, to combine harvester, laser or robot. Have you got the right type? We will look at the two main types: the 'cylinder lawnmower' and the 'rotary grasscutter'. The cylinder machine with its cylindrical cutter, cuts like scissors, whereas the rotary machine, which has a propeller-like blade, cuts by thrashing the grass. People ask, 'Which is the best type?' The answer is, both are equally good, but do different jobs and give different finishes. Each has its own advantages and disadvantages; the secret is choosing the right one for you and your grass. Getting this right is one of the secrets to saving you money, work and hours of frustration.

When purchasing a new machine it is well worth going to a garden machine specialist or outdoor power equipment dealer and ask for advice – it's what they are there for. Feel the weight and the balance of the machine: it may look good in an advert but may be a lot heavier or lighter than you thought. When you get hold of it you can make an educated and sensible decision about which one to choose.

You can compare the purchase of a lawnmower with a new car: the price goes up with the more power, size and extras, but this is not necessarily a good or bad thing. Ask yourself: do I need a Rolls Royce just to go to the local corner shop? Conversely it's no good buying a Mini car if you need to travel two hundred miles a day on the motorway. It's very relaxing to travel long distances in the Rolls Royce with no effort, but awkward to park it in a tight spot in town on a busy day. The Mini's great for parking and man-

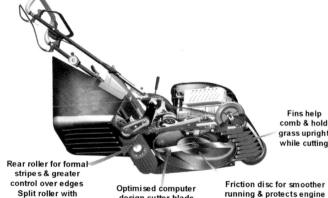

The latest Hayter Harrier (x-ray view) features an aluminium chassis, rear roller, variable walking speed, impact damage protection and height adjustment control via a single lever.

Rear roller for formal stripes & greater control over edges Split roller with differential gear for easy turning

Optimised computer design cutter blade for low noise, vibration & better collection

Fins help comb & hold grass upright while cutting

Friction disc for smoother running & protects engine crankshaft from accidental impact damage

oeuvring but it's hard work travelling long distances with its engine working flat out. Mowers are just the same: firstly a machine with a wider cut and more power will do the job more quickly. This means you will go up and down the lawn fewer times; also wider machines have a larger capacity grassbox, which equates to fewer trips to the compost heap. Overall this means a faster job with both you and the machine working fewer miles and hours, and generally the machine will last much longer. However, a quality hand-powered lawnmower may well be quicker and easier on a small lawn – you can have the lawn cut and finished by the time you've filled a power mower with petrol, checked the oil and started it, or on an electric mower unravelled the mains cable, found the extension, plugged it in, cut the grass then stored it away again.

What finish do I require?

If you just want to keep the grass neat and tidy

ABOVE: A cylinder petrol lawnmower is ideal for short grass. This machine is designed to achieve a formal striped close cut, the perfect British traditional lawn finish.

RIGHT: The Hayter Harrier 41, a rotary mower with rear roller. It rolls and stripes, will cut reasonably long and short grass without dropping off the lawn edge, and is therefore a good compromise between a cylinder and a four-wheel rotary mower.

LEFT: A four-wheel rotary mower is ideal for longer grass, simply achieving a neat and tidy finish.

the quickest and easiest way, a four-wheel rotary mower would be the machine best suited for you. Conversely, if you want to cut the daisies with a close cut, and a rolled and striped traditional finish right to the edge, a cylinder lawnmower with roller would be the better option. As a compromise, a rotary with a rear roller will give you traditional stripes and cut the grass shorter than a four-wheel rotary mower.

Consider what sort of lawn you have got. Be honest with yourself: everyone would like that perfect bowling green finish, but in reality many people know that after a while it will probably be neglected, and it's not going to be weeded, fed or loved. Our social habits are changing – traditionally many people used to cut their lawns on a quiet Sunday; now on Sundays we have the option to go shopping. Or we could leave the robot mower cutting the grass at midnight, and go out shopping when we would normally be sleeping. If you want your lawn to be as near to a bowling green finish as possible, a cylinder-type lawnmower will achieve it. It cuts like scissors, giving a clean, perfect cut, which allows the grass to heal more quickly. Models vary from having three blades up to twelve blades – the more blades, the finer the finished cut. Cylinder lawnmowers cut in one direction only, which gives a much more defined stripe, as opposed to a rotary cutter that cuts in all directions. If all you want is to keep the grass neat and tidy stick to the rotary type. The rotary thrashes the grass like a scythe, tackling longer and rougher grass areas.

Besides a good lawnmower, to get that truly envious superb show lawn, you need to feed it, weed it and love it, as once the grass becomes more than few inches in length it ceases to become a lawn and becomes grass, which then requires a grasscutter. Cutting 'little and often' is one of the golden rules, ideally more than once a week. Also, very importantly, never cut more than a third of the grass height at any one time.

CALCULATING THE SIZE OF LAWNMOWER REQUIRED

An easy way to work out the correct size is to compare your lawn to the size of a tennis court. A tennis court is approximately 78ft × 36ft, i.e. 312 square yards (approx 24m × 11m, i.e. 264 m²).

Up to half a tennis court: 12in to 16in (30cm to 40cm) width of cut.

Up to a tennis court: 17in to 21in (42cm to 52cm) width of cut.

Up to two tennis courts: 21in to 26in (53cm to 65cm) width of cut.

What size do I need?

Choosing the correct width of cut is one of the keys for saving you time, as the wider the cut, the less time you have to spend travelling up and down. The fewer the journeys up and down the lawn (and the fewer the trips to the compost heap), the fewer the hours worked, so the machine will last longer. Many people who find cutting the grass a chore have chosen too small a machine.

Most traditional cylinder lawnmowers with full metal rear rollers have the grassbox mounted on the front of the machine (although some modern cylinder side wheel and light electric models have the grassbox fitted at the back), with the grass-cuttings being thrown in a forward direction away from you. The advantage of this is that you can see exactly how much grass you are cutting, how it's cutting and exactly how full the grassbox is. There is also no obstruction like the handles and cables when emptying the box. The disadvantage is that it is harder to get right into corners without flattening the flowerbeds – unless the grassbox is removed. On rotary grasscutters the grassbox is fitted on the back of the machine with the grass being thrown backwards. Rotary mowers with a rear roller have the advantage of being able to cut much nearer to a corner without flattening the flowerbed. The blade also comes nearer to the edge of the machine, enabling it to cut closer to a wall, tree or post, unlike some four-wheel rotary machines; with these it can be difficult to cut the border without the wheels dropping off the lawn edge. The dis-

CHOOSE THE RIGHT MOWER FOR THE RIGHT FINISH

- Basic grass cutter

Hover mower

Hand mower
Side wheel hand mower

- Neat and tidy
- Good grass collection

4-wheel rotary mower

- Rolled and striped
- Cut to lawn edge

Rear-roller rotary or cylinder lawnmower

- Rolled and striped
- Cut to lawn edge
- Groomed lawn finish
- Lawn raked and scarified

5- or 6-blade cylinder lawnmower

Cylinder hand mower

- Rolled and striped
- Cut to lawn edge
- Ultimate bowling green finish

10–12-blade cylinder lawnmower

advantage of the grassbox at the back is that as it is enclosed (without its safety flap you would get covered in grass) it is harder to tell when the grassbox is full, and if it is overfilled it will clog the machine.

When purchasing a new mower its worth checking the grassbox capacity size, how convenient and easy it is to remove and attach, and how easy it is to tip and empty the grassbox. One other aspect to check is how easy it is to adjust the height of cut. This can vary from getting under the machine and undoing nuts and bolts, to four separate adjusters (one on each wheel), to a single adjuster or knob that alters the whole machine all in one movement. These adjustment levers vary enormously: some are easy; some are not quite as user-friendly. Check the height adjustment control yourself or ask for it to be demonstrated before purchasing your machine. It can be the difference between what should be a simple few seconds of a job, or a frustrating irritation that you have to get the neighbour to help you with. Cylinder lawnmowers have a much shorter wheelbase than a rotary, so on uneven surfaces they are less likely to scalp the lawn when set on a close cut.

What type of power?

Lawnmowers and grasscutters can be petrol-powered, mains- or battery-powered, or hand-powered. Each method has its own advantages and disadvantages. If the lawn is tennis-court size or larger, petrol power would probably be most suitable, since a mains cable in this instance may become too inconvenient. All gardens are different so take into consideration any trees, ponds, flowerbeds etc. that the mains cable may have to negotiate. Also remember that the nearest power supply may be a fair distance from the lawn. If the cable is more than 50 meters long you may start to get a power drop, which can cause the motor to labour and overheat. A cordless machine would be worth considering if the garden has many obstacles and awkward shapes. Battery technology has become smaller, lighter and more powerful since 2006 with the introduction of Lithium Ion batteries, which can recharge more quickly. Battery machines have the advantage of having no petrol, no oil, no fumes, and no starting issues; in addition they require less maintenance and are generally quieter than their petrol counterparts. Disadvantages are that you can run out of power before finishing (purchasing an extra battery will double the cutting time available, and some models are designed to continue to cut even when the battery is flat).

The latest technology is the robot mower: although first introduced in the late 1950s, petrol-driven robots were not properly marketed until 1995. These new machines are manufactured with solar, battery or petrol power. Once set up the robot will trim the grass without attention when and where you want, day or night. (For more information on robot mowers, *see* Chapter 10.)

The hand-powered cylinder lawnmower has remained virtually unchanged over the last 180 years since its invention, although the materials are now much lighter. A good-quality hand push lawnmower surprisingly might initially cost more than a power mower but in the long term it can last a lot longer and with far less maintenance cost. Besides being quieter and giving an excellent finish, they are also the most environmentally friendly, with no carbon emissions and using the minimum of materials. A lot of these machines, especially the older ones, will last more than a lifetime if well maintained. Hand-powered lawnmowers are easy to use provided they are sharp and the grass is short. Cylinder machines come in various sizes from 6in to 30in (15cm to 75cm). Rotary machines range from 10in to 54in (28cm to 136cm).

Stainless-steel mulch mower with non-rust stainless-steel chassis. The handle, which moves from side to side, allows you to avoid walking in the bushes and where you have mown. Mulch mowers can save up to a third of the cutting time and an average of 1.25 tons of collected grass every year.

Grass collection

There is one other consideration: to collect or not to collect the grass? The advantage of collecting is that it will give the best finish and provide compost for the rest of the garden. The disadvantage: it will take more time to cut and you will require a compost heap, and in the case of a big lawn, a big compost heap. One of the advantages of not collecting the grass is speed, especially if it's long. However, the cut grass may end up in clumps and when it dries will look unsightly. There is an alternative solution if you do not want to collect or have unsightly clumps of grass left all over the lawn: this is to use a mower that will cut the grass into fine pieces, called a 'mulch mower'. These mulching machines are designed to cut the grass into a much smaller volume. The blade holds the grass inside the cutting deck much longer, chopping the grass into small particles; the grass is then dropped onto the ground and decomposes very quickly, leaving virtually no grass visible after cutting. The advantage of mulching has several other benefits: the finely chopped grass acts as a fertilizer to feed the grass roots, leaving it quite lush and providing about a quarter of a lawn's annual needs (if you haven't fed your lawn for a long time it will think it's its birthday). Mulching is especially beneficial when the weather is in drought conditions or is hot and dry, since the clippings are about 80 per cent water. Mulching slows evaporation from the soil surface, and conserves moisture.

The grass mulch preserves a lot more moisture, stopping the grass from burning and going that horrible yellow colour in a drought. Another benefit is you don't need a compost heap, saving on average 1.25 tons of collected grass every year on a lawn the size of an average tennis court. Dedicated mulching machines will even mulch leaves (providing your lawn is not in the middle of a forest). They are also good for the environment particularly if you are currently binning your grasscuttings. For the best of both worlds some mowers will collect and mulch; these machines are often called '3 in 1: Mulch, Collect or Drop' but they can sometimes be a compromise, not giving quite as good a performance as a dedicated mulcher or a dedicated collector. The secret of mulching is to cut 'little and often'.

CHAPTER THREE

LAWNMOWER ENGINES: PRINCIPLES AND PURPOSES

There are two common types of petrol engines fitted to garden machinery: four-stroke and two-stroke engines. Each type can have either a horizontal crankshaft for a lawnmower (like a car), or a vertical crankshaft for a rotary mower (like a helicopter). In the past, two-stroke engines have mainly been fitted on hand-held machines because of their lightweight advantage, but recently modern technology has been able to reduce the weight on four-stroke engines. We will see more of these lightweight four-strokes in the future as new technology and strong, lightweight materials become more cost-effective.

The crankshaft is the main part of any engine and supplies the direction and the drive; it needs good clean oil lubricant all the time for optimum engine life. Once the crankshaft is worn (normally through lack of oil or contaminated old oil) in most domestic mower engines it can be terminal and the mower destined for the Great Mower Scrap Yard in the Sky, as in many circumstances it can be uneconomic to replace. Especially in the case of budget mowers there may not be a great deal of difference between the cost of replacing a crankshaft, engine or complete mower. (As the interviewer asked the pilot on the first solo flight across the Atlantic: 'Would there be a problem if one of the engines failed?' 'Yes – it's only got the one!')

Four-stroke engines have the same basic design as that used in cars. Each type of engine has its advantages and disadvantages; each is suited for different jobs. The majority of four-stroke engines are heavier than two-strokes, so most are fitted to machines with wheels or rollers where weight is not a problem and where more power is required. They are ideal for lawnmowers, garden tractors, cultivators, chippers, shredders, generators, etc. Two-stroke engines are much lighter as they have fewer moving parts; they also have the advantage of being able to run at any angle and they are ideal for steep slopes (they do not have an oil sump as the lubricating oil is mixed with the petrol). Two-strokes are ideally suited to hand-held machines like chainsaws, brush cutters, hedge trimmers, grass trimmers, hover mowers, and also boats and motorbikes. Each type may run on a different fuel mixture, which is important especially in two-strokes: *run a two-stroke engine with incorrect fuel mix and it can result in permanent engine damage within seconds.*

So, to check which type you have: all four-stroke mower engines require oil in the sump just like a car. The oil's purpose is to cool, seal, clean and lubricate. Look for an oil dipstick or oil screw plug; if there is one, it is a four-stroke engine. Most of these oil caps can always be seen from the top of the engine. It is well worth getting into the habit of always checking the oil level before starting the engine: if the oil is low, top it up; if the machine has been run for more than twenty-five hours, change it. If the engine is brand new,

THE FOUR CYCLES OF THE FOUR STROKE ENGINE

Intake stroke. As the piston moves down the cylinder, a vacuum occurs. The intake valve opens via a cam gear. Atmospheric pressure pushes fuel/air mixture into the cylinder above the piston. When the piston is at the bottom of the stroke the intake valve closes. The exhaust valve stays closed.

Compression stroke. The piston moves up the cylinder with both valves closed, highly compressing the fuel.

Power stroke. Just before the compression stroke ends, the magneto produces a high voltage arc across the sparkplug gap, igniting the fuel, which produces the pressure to push the piston down.

Exhaust stroke. As the piston moves up, the exhaust valve opens via the cam gear, and the piston pushes the burned gases out, completing the fourth cycle where the first stroke begins again.

WARNING. Exhaust gases contain carbon monoxide, which is odourless and poisonous. Ensure care is taken to provide adequate ventilation when running an engine in an enclosed area.

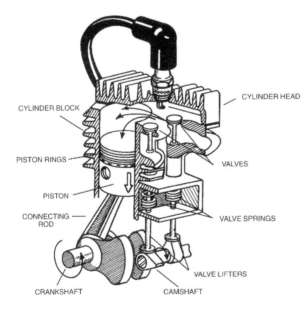

LEFT: A four-stroke engine, cut-away illustration.

CYLINDER HEAD

CYLINDER BLOCK

PISTON RINGS

VALVES

PISTON

CONNECTING ROD

VALVE SPRINGS

VALVE LIFTERS

CRANKSHAFT

CAMSHAFT

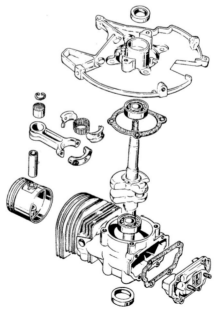

RIGHT: A two-stroke engine. Note there are fewer moving parts compared to a four-stroke. Two-stroke engines use port channels to convey fuel and exhaust gases instead of valves as in a four-stroke.

The four strokes of the four-stroke engine.

a. INTAKE STROKE: The piston goes down, creating a vacuum in the cylinder which draws gas through open intake valve into the space above piston.

b. COMPRESSION STROKE: The piston comes up with both valves closed, highly compressing the gas into the space left between the top of the piston and cylinder head.

c. POWER STROKE: At this point the magneto sends high tension current to the spark plug, firing or exploding the compressed gas and driving the piston down.

d. EXHAUST STROKE: Exhaust valve opens and the upward stroke of the piston forces out all of the burnt gases, thus completing the power cycle.

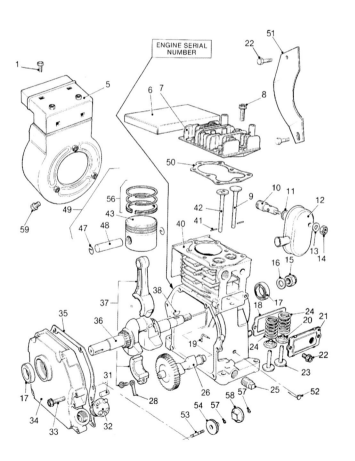

Illus. No.	Description
1	Bolts for cowl
5	Cowl sub assembly
6	Gasket set
7	Cylinder head
8	Bolt - cylinder head
9	Inlet valve
10	Connecting pipe
11	Washer gasket
12	Silencer sub assembly
13	Washer for silencer
14	Nut for silencer
15	Nut for crankshaft
16	Washer for crankshaft
17	Oil seal
18	Breather gasket
19	Bobbin follower
20	Valve spring retainer
21	Breather assembly
22	Screw for breather & front cover plate
23	Tappet
24	Valve spring
25	Drain plug
26	Camshaft
28	Oil splasher
31	Dowel pin
32	Oil filler plug
33	Screw for crankcase cover
34	Crankcase cover sub-assy.(includes No.53)
35	Gasket - crankcase cover
36	Crankshaft
37	Connecting rod assembly
38	Key for flywheel
40	Cylinder block
41	Cotter pin valves
42	Exhaust valve
43	Piston
47	Wire circlip
48	Gudgeon pin
49	Piston assembly
50	Cylinder head gasket
51	Front cover plate
52	Ratchet plate
53	Spindle (not supplied separately)
54	Gearwheel assembly
56	Piston ring set
57	Circlip
58	Bobbin
59	Screws for cowl

Atco A114 (Suffolk 2hp engine), exploded view. This typical four-stroke engine is fitted on many Suffolk, Qualcast, Atco and Webb lawnmowers and cultivators.

change the oil after five or six hours' running time. For the correct oil grade and capacity check your owner's handbook or contact your local service dealer.

New and efficient four-stroke engines are now coming onto the market. These are more environmentally friendly, lightweight, fuel-efficient, quieter, and can be run at any angle. At present, environmental issues are changing and small-engine technology is changing to match it. The correct fuel is becoming much more important than in the past, as modern engines are becoming more efficient and less polluting. (There are exceptions: some engines, including Stihl products, are four-stroke with no sump or dipstick, which run on two-stroke fuel. Currently these are mainly fitted to hand-held machines like brush cutters, grasscutters, long hedge cutters, chainsaws, etc.) Modern four-stroke engines run on plain unleaded petrol; the lubrication oil is put in the sump, which is normally filled to a dipstick with minimum and maximum marks – or if there is no dipstick and it is fitted with only a screw cap, oil is normally filled level with the top of the filler hole.

Oil

Make sure when filling engines with oil that the engine is on a level surface. If the engine is run too low in oil it will overheat and prematurely wear all the moving parts; this is the main cause of premature engine wear. Also if the engine is overfilled with oil it will not run properly. The correct oil grade and quantity are very important if you are going to get the best reliability and the longest life out of your engine. The correct oil grade for the engine is sometimes printed on the machine; if not, it will be in the owner's instruction manual. If this is not to hand, consult your garden machine dealer, quoting the engine model, and they will advise you which oil to use (especially if you are not sure if you have a four-stroke or two-stroke engine).

Recommended oil manufacturers (from left): Honda (1l – 08221-888-101HE); Briggs & Stratton (0.6l – 992997 for all walk-behind machines, 1.4l – 992996 for 7hp engines and above); Mountfield (1l – MX855); Atco/Qualcast/Tecumseh (1l – F01610258).

Oil change kit, containing syringe, pipe and bottle: an easy and clean method of emptying the engine sump without tipping the machine up, without the need for spanners and without spilling any oil.

However new the machine, if the engine is run starved of oil or if incorrect fuel is used, it will not be covered under the manufacturer's warranty and can end up requiring a costly repair. There is no harm in asking a garden machine service specialist for advice on which is the correct oil to use. (Most garden machinery manufacturers make only the chassis and do not make any engines, so if any parts are required for the engine always quote the engine make and model as well as the chassis make and model.) All engines fitted on garden machinery are air-cooled and run at different temperatures to car engines, which are water-cooled, therefore different oil grades are required. If a lawnmower is run on a hot day under hard working conditions, the engine will run hotter, and the quality of lubrication of the engine oil can deteriorate within an hour. If an incorrect oil is used (even an expensive oil) it will burn off a lot of the oil's additives and lubricating qualities, whereas in a water-cooled engine like a car the water will always keep the engine temperature at the correct level without overheating. In general, popular domestic four-stroke Briggs & Stratton, Qualcast and Suffolk lawnmower engines use straight SAE 30 grade oil. A straight 30-grade oil is all oil and nothing but the oil, unlike multi-grade oils, which have additives that can burn off if the engine overheats, especially if run in heavy and hot conditions. Briggs & Stratton and Qualcast manufacture a heavy-duty grade ideally suited to their engines; this is inexpensive and well worth it. Some come in a bottle with the exact quantity for your engine so there is no guessing if you've put in too little or too much. Honda mower engines recommend SAE 10W-30 Mountfield 10w/40 multi-grade oil, and have a dedicated oil recommended for their engines.

The best time to drain engine oil is when the engine is warm. The oil is thinner when warm and the oil and sludge will flow out more easily; it is less likely to leave anything in the bottom of the engine's sump. There are several ways you can drain the oil (always check your specific engine handbook).

Most horizontal engine crankshaft types, especially those fitted on cylinder lawnmowers, are the easiest. Drain the petrol or fuel by running the engine. Many have a sump plug, bolt or screw either at the front or back, at the base of the engine. (Some of these engines may be fitted with two filler plugs: depending on the engine application, either plug will do.) Slacken off the most accessible plug (if possible choose the bolt nearest to the handles as it makes it easier to tip the machine backwards), place a small bowl or suitable container under the plug. You may need to tilt the machine slightly toward the bowl for a clean flow, and then remove the plug, let the oil drain till it stops dripping. (This can take several minutes, and an extra pair of hands can be useful at this stage.) Be careful that the engine is not too hot: the oil will be hot so wear gloves. The oil may also come rushing out, so be ready for the flow; oil makes a mess if it's spilt.

On vertical crank engines changing the oil is different. To empty the oil, most machines can be tipped sideways, but consult the owner's handbook or your local service dealer to establish which way it can be tipped without contaminating the carburettor, air filter, silencer or cylinder head with oil. Many vertical crank engines have a sump plug or nut underneath the engine – often it can be awkward to access especially if a power drive or blade clutch disc is fitted. First of all, empty the fuel. Before tipping, always remove the spark plug cap. Pull the rope slowly to the compression stroke (when the ropes goes stiff), which helps to prevent oil from entering the cylinder head.

The easier alternative is to use a simple oil change kit. These inexpensive kits are quick, and eliminate the need for tipping the machine. No tools are required, no rapped knuckles, and the chances of any possible oil spillage are reduced. The kit comprises a bottle and a syringe. Simply

Aspen cartoon advertising alkylate fuel. Ever had a headache whilst using petrol from the pump?

remove the oil filler plug or dipstick, place the syringe inside the filler hole to the base of the engine, and suck the oil out. Then squirt the oil from the syringe straight into a suitable container. It may not be quite as thorough as a conventional oil change but if it means the difference between changing or not changing the oil (as is mostly the case) the oil change kit is an excellent option. Oil change kits can be obtained from The British Lawnmower Museum or any

---

GENERAL RECOMMENDED OIL GRADES FOR DOMESTIC FOUR-STROKE ENGINES

Note: capacities and grades can vary so always check the owner's handbook for your particular engine. Always check the oil level after filling: too much or too little oil can damage the engine.

Briggs & Stratton	SAE 30 API SF/CC	0.6–1.4 litre
Briggs V Twin	SAE 30 API SF/CC	2.0 litre (with oil filter)
Suffolk	SAE 30 API SF	0.5 pint
Qualcast	SAE 30 API SF	0.5–1 pint
Kohler	SAE 30 API SF	1.4 litre upward
Tecumseh	SAE 30 API SF	1 pint
Aspera	SAE 30 API SF	1 pint
Villiers	SAE 30 API SF	0.5–1 pint
BSA	SAE 30 API SF	0.5–1 pint
MAG	SAE 30 API SF	1 pint
Honda	10w/30w API/SJ	0.6 litre upward
Mountfield	10w/40w API SG/CF	0.6–1.4litre
Kawasaki	10w/40w	1 pint upward

RECOMMENDED OIL MIXTURES FOR DOMESTIC TWO-STROKE ENGINES

Note: petrol evaporates; oil does not. It is therefore advisable not to store two-stroke petrol mixture for long periods as the fuel will end up with the wrong oil ratio.

Flymo Hover	25:1	Flymo/Husqvarna oil
Tecumseh Hover	25:1	Flymo/Husqvarna oil
Aspera	25:1	Flymo/Husqvarna oil
Stihl	50:1	Stihl oil
Husqvarna	50:1	Husqvarna oi
McCulloch	40:1	McCulloch oil
Partner	50:1	Husqvarna oil
Ryobi	50:1	Ryobi oil
Flymo	50:1	Flymo/Husqvarna oil

good service dealer. If in doubt, always consult your owner's handbook.

Another handy, quick and easy oil change device is a small pump, which fits on the end of an electric drill. Firstly drain the fuel by running the engine until it stops, then place the pipe in the engine oil filler tube and the other end into a suitable container like a bottle or a can, start the drill and, hey presto, the oil will be automatically sucked out, as there is no need to prime the pump. This device is especially useful on machines with larger oil sumps, or those that are hard to tip over, or for machines that have awkward oil drain plug access, as fitted on some garden tractors. There are several makes of pump on the market which look similar, but check that the pump is suitable for oil (for example oil evacuation pump Briggs & Stratton Part number 5056D). Do not use this type of pump to evacuate petrol, as fuel vapour may possibly ignite from a spark from the drill.

Note: these are general grades, specifications and capacities, which can alter depending on the model engine, always check in your handbook. If you don't have a handbook ask your garden machinery dealer for one. (There may be a small charge but it is well worth it.)

Fuel

The fuel is just as important as the oil if your engine is to keep running smoothly for an extended lifetime, with reliable starting. Modern fuels from the petrol stations are mainly designed for cars, which have much more sophisticated engine management systems compared to some of the very basic domestic mower engines. (Remember some mowers are made with serious economic restraint and have been designed to the lowest cost.)

Petrol can also vary enormously from garage to garage with so many different companies and sources blending fuel from all over the world. Modern petrol, to pass the current fuel emission standards, has had many process changes. This has the advantage of not polluting the atmosphere as much, with the disadvantage that the fuel does not last very long – in some cases as little as a few weeks.

Alkylate fuel

Alkylate fuel is the cleanest of all the petrol types. Originally intended for professional users twenty years ago, it is now widely being used in a growing number of environmentally aware companies and organizations that are making a change, protecting the health of the users, those around, and the environment, such as the National Trust, London Zoo, Whipsnade Zoo, Longleat, various councils, and Scottish and Southern Energy. Fire and rescue services use it for greater reliability for starting equipment – especially important in life-or-death situations. It has many big advantages over petrol that has been bought from the pump. Alkylate fuel can remedy and eliminate a lot of engine problems such as sooty spark plugs, piston deposits, and hot start issues; it leaves your engine running smoothly for a much longer service life with the advantage of not having as many trips to the service department. If you want to avoid the frustration of an engine not starting after winter, or if you want to store your engine for long periods without draining the fuel each time, alkylate fuel may be the answer; it can even clean away the deposits in engines originally run on pump fuel

Aspen 2T and 4T alkylate fuel in 1l and 5l containers.

from the petrol station. (Smoking may occur on change-over for a short period.) The environment will also benefit from a reduction in carbon monoxide emissions, and plant life is also not destroyed by ground ozone.

Dirt in the fuel is one of the most common engine problems, very often brought back from the petrol station: the fuel is stored in large metal tanks where condensation can form, causing droplets of water and rust. In the case of cars which have much better filters and larger fuel jets it is not a problem, but a blockage in the much smaller jets in a mower or a chainsaw leads to poor engine performance and starting. To keep the cost of fuel down, more petrol companies are now adding ethanol (alcohol) to pump fuel. Ethanol can attract moisture: if the carburettor and tank have been cleaned of water in a service, and then within a few months the problem recurs, the cause could be the ethanol in the fuel attracting more moisture. Too much ethanol will also start to erode plastic engine parts especially in the mower carburettor and fuel tanks. Engines run on alkylate fuel will eliminate this problem completely, especially if alkalyte fuel is used from the start.

Secondly, petrol from the pump deteriorates in as little as four weeks. In the case of two-stroke engines when oil is mixed with the fuel the deterioration can occur within only a few weeks. Aspen alkylate petrol lasts for years without any loss of quality – in fact up to three or four years – so no fuel has to be thrown away. If your fuel is a few months old and you have not added a fuel stabilizer, the environment agency recommends putting it into your car petrol tank as long as it is at least half full; then it will mix harmlessly. (Or you could ask your local fire station if they would like your old mower petrol for fire practice.) Alkylate two-stroke fuel does not require the mixing of oil and petrol: a great advantage since it eliminates potentially getting the mix wrong and saves time.

Thirdly, alkylate fuel keeps valves and pistons cleaner, giving a longer engine service life and greater operation reliability. Fourthly, with alkylate fuel there are reduced fumes from the exhaust and unburnt fuel mixture; this is a particular advantage with hand-held machines such as chainsaws, hedge cutters and grass trimmers, where you are operating the machine much closer to the exhaust gases. Fifthly, any fuel spills mainly evaporate (unlike fuel from the pump); the oil is also biodegradable, and even the polyethylene plastic container can be recycled; during incineration the only substances formed are carbon dioxide and water.

Finally (actually a point unrelated to mowers), alkylate fuel is great for camping and fishing and can be used in Coleman dual fuel camping stoves, as it is a lot cheaper – currently at nearly half the price. And for all those people cleaning out engine parts, mower bits, carburettors (even guns) etc. with petrol in the kitchen sink (which we do not recommend) to the great displeasure of the wife, where normally the foul petrol smell would lead to divorce, the problem is solved when cleaning with alkalyte fuel.

If you're still not sure about this fuel and are sceptical about its efficacy, go into your garden shed and take a quick sniff from your petrol can, then pop down to your local garden machinery specialist and ask if you can just have a sniff of alkylate fuel. You will be in for a surprise: it looks very clear, almost like water, and there is virtually no smell – and what slight smell there is, is quite pleasant.

Alkylate four-stroke (4T) and two-stroke (2T) petrol was designed initially for health reasons in cooperation with professional users. It contains virtually no harmful substances such as benzene, lead, sulphur or aromatics. It is sold in 1-litre or 5-litre bottles, 25-litre cans and 200-litre drums.

Aspen 4T (four-stroke) is for lawnmowers, cultivators, chippers, snow blowers, aerators, boats, etc. Aspen 2T (two-stroke) is for chainsaws, hedge trimmers, clearing saws, brick saws, some mopeds, and many other air-cooled hand-held machines that require oil and petrol mixture.

Fuel stabilizer

The alternative to alkylate petrol is to add a fuel stabilizer to pump petrol. There are many brands on the market to choose from. Fuel stabilizer keeps the petrol stable at whatever the octane grade it was when you added it. The best time to add fuel stabilizer is as soon as you buy petrol. Remember modern fuel starts deteriorating after four weeks, and it is quite possible the petrol in the pump may already be two or three weeks old before you purchase it; within another couple of weeks it will start to lose its volatility. After several months it will start gumming up the carburettor and engine with a varnish-like substance, and if left over the winter the fuel will have a low volatility, which may cause poor starting or will run the engine at a higher temperature, causing premature wear.

A common fuel stabilizer called Fuel Fit is made by Briggs & Stratton. It comes in several sizes, starting with a 30ml bottle, which will stabilize up to 30 litres (6 gallons), and 125ml, which will stabilize up to 125 litres (25 gallons) of petrol. It is cost effective and inexpensive, and will keep petrol fresh and stable for up to two years. Fuel stabilizer eliminates some temperamental engine problems, promotes quicker starting, keeps carburettor and fuel clean, removes gum and varnish build-up and improves engine performance. If installing fuel stabilizer just for the winter or longer, run the engine for a short period to circulate the additive through the carburettor and engine. There is also a fresh start cartridge that just fits in the engine's petrol filler cap – this is ideal for all Briggs & Stratton Quantum engines or any engine with the same fuel cap size. Ensure the fuel is fresh when using additive and the

Briggs & Stratton Fuel Fit stabilizer for four-
and two-stroke engines. (30ml –992999;
125ml – 999005E.)

engine and fuel can be stored for up to twenty-four months. Fuel stabilizer itself also has a very long shelf life.

Tips for winter storage

To ensure your two- or four-stroke engine is going to start after the winter period, after your last cut, remove the spark-plug and pour 15ml (0.5 fl oz) of clean engine oil into the cylinder bore. Replace the spark plug and crank the engine over slowly to distribute the oil; this coats all the internal metal parts and stops rust forming.

Another winter tip for four-stroke engines is to leave the engine on the compression stroke. When the engine is left in this position it will prevent the valves from seizing open, eliminating a no-start for the first pull of the season. To achieve this, pull the starting rope slowly till the rope gives a resistance; as soon as it gets to this point stop pulling. If you are not sure you are at the point of most resistance, keep pulling slowly till you feel the next resistance on the rope. At this point the piston is at the top of the engine, which means that both the inlet and the outlet valves are in the closed position. When the temperature drops, less condensation will occur on the engine bore. Condensation causes droplets of moisture to form on the metal, which can cause a thin film of rust on the valve stems; in the compression position the valves will not seize open over the winter, and it will run free when you start it up in the spring, even if the machine has been left in the cold and damp without being run for several months.

If you were to leave your car outside for six months over the winter without touching it, it probably wouldn't start, so you shouldn't expect too much of the mower in March or April if it was left exactly where it cut its last patch of grass in October. Try to keep your mower in as dry a place as possible. It's also worth running the engine every four or five weeks just for about five minutes or so to give the engine internals a coat of oil; this will also dispel any moisture in the engine when it warms up.

If during the cutting season you have not been using a petrol additive preserver or alkylate fuel, the last cut of the season is the one time to use it. By using this fuel there is no need to empty the fuel tank; also it will stop the fuel gumming up any of the carburettor jets. Petrol straight from the petrol station pumps or untreated fuel over several months starts to turn into a varnish-like consistency and the octane rate dramatically drops, especially if it has been left in a can exposed to air. The warmer the fuel and the more air in the container or tank, the quicker the fuel will deteriorate. Petrol bought in the winter has a higher volatility than in the summer.

QUICK DIAGNOSIS ON COMMON ENGINE PROBLEMS

Engine won't start

1. *Electric problem:* spark plug, ignition, cut-out switch or flywheel key.
    Remedy: replace spark plug, check plug lead, cut-out switch, clean points, check coil and flywheel key.
2. *Fuel problem:* no fuel, stale or contaminated fuel, blocked fuel pipe/tap/carburettor, engine flooded, or clogged or unvented fuel cap.
    Remedy: if the spark plug is dry, dip the plug tip in fuel then refit; if the engine starts, the carburettor requires attention.
3. *Compression problem:* valves, piston or connecting rod.
    Remedy: will possibly require an engine strip down – take to a professional.

Engine runs rough

1. *Engine misfire:* spark plug, points, carburettor setting, valves or clogged filters.
    Remedy: change spark plug, clean points, readjust carburettor, check ignition coil and filter.
2. *Engine overheats:* low oil level, blocked cooling fins, lean fuel mixture.
    Remedy: top up or change the oil, clean engine and flywheel cooling fins, check in tank for any fuel contamination, change fuel if older than six weeks.
3. *Engine smokes:* rich fuel mixture, clogged air filter, overfilled with oil, machine has been tipped.
    Remedy: check oil level, change fuel, change or clean air filter. If machine has been tipped and smoking does not stop within five minutes take to a professional.
4. *Engine knock:* loose flywheel, loose blade boss, excess carbon in combustion chamber.
    Remedy: tighten blade boss, check flywheel key, de-coke engine, check fresh fuel or replace.

CHAPTER FOUR

MODERN CYLINDER LAWNMOWERS

The cylinder: cutter and height adjustment

The cylinder cutter (the part that spins rapidly at thousands of revolutions per minute) is made up of between three and twelve curved blades – the more blades, the finer the cut. A basic five-blade hand mower will give between thirty-four and thirty-six cuts per yard. A six-bladed hand mower will produce forty-five cuts, and an eight-bladed hand mower sixty cuts per yard. Six-blade petrol lawnmowers can achieve over a hundred cuts per yard depending on the model. Twelve-blade cutters are for bowling or putting greens, which can achieve over 150 cuts per yard.

The blades are fixed to the cylinder shaft by three to six 'spiders', depending on the cutter width. The cylinder shaft is fitted into each side of the chassis side plates on ball bearings, either on a cup and cone bearing or a ball-race, often the self-aligning type. Cup and cone bearings have a single row of balls held apart by a lightweight cage that run around a cone shape shaft; they are normally less expensive than standard ball races, which consist of a larger quantity of balls with a single or double row housed within two precise machined rings. The cylinder needs to spin freely and smoothly; if the cutter is stiff to turn, the bearings need to be greased, cleaned or replaced. Although they are shielded to prevent grass and water ingress, eventually the bearing will be contaminated by some type of unwanted material. The cylinder is positioned directly above the bottom blade, which is attached to a sole plate with screws, bolts or rivets. The sole plate is also often referred to as the bottom plate, sole-

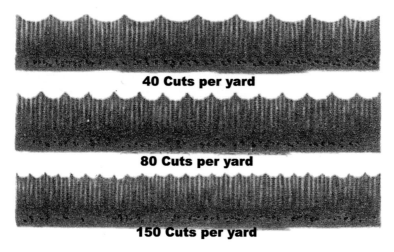

40 Cuts per yard

80 Cuts per yard

150 Cuts per yard

Forty cuts per yard: four to five blades; eighty cuts per yard: six to ten blades; 150 cuts per yard: twelve blades. The illustration shows that the more blades, the smoother and the finer the finished cut.

plate casting, cutter bar, bottom blade and knife-bed. Bottom blades can either be separate or integral with the soleplate.

On cylinder lawnmowers, the cutter is often adjusted incorrectly – it is the single main cause of poor cutting on any cylinder lawnmower. The symptom of this is a laddered effect on the grass. Either the cutters are set to not cutting at all (called 'off set') or set too tightly (called 'too hard on'). When the cutters are set too hard on, it causes the cylinder blades and the bottom blade to wear prematurely: the harder on, the more friction is caused, and the more the blades will heat up and wear. There is also considerably more strain on all the transmission, gears, drive belts and chains, plus a noise that sets your teeth on edge. On an electric mower it can possibly cause the motor to overheat and burn out, espe-cially if coupled with cutting long grass at the same time. If the cutters are off set, there will be no detrimental damage to the machine, just the annoying frustration of simply not cutting the grass! There may only be part of the blades con-tacting the bottom blade; in this case the cutter is out of true, possibly caused by an impact or it could be a worn bearing. Alternatively, the blades could just be blunt or dull. Both these problems would be best dealt with by a professional who can offer a regrind on a lathe.

On cylinder lawnmowers there are two types of cutter adjustments, which are often and easily confused. One type is to adjust the cutter setting for the 'keenness of cut' – this is done by either moving the cylinder up or down onto a fixed bottom blade, or moving the bottom blade up or down onto a fixed cutting cylinder. The other is to adjust the cutter for the 'height of cut', which is always altered via the front rollers being moved up or down. Both may require attention during the cutting season. The principles on adjustment for the cutter and height settings have not changed for the past 180 years, from 1830 to the present day. A cylinder cutter works

Cutting cylinder – screws for adjusting the keenness of cut.

Height of cut adjustment via the front rollers. (Keenness of cut is adjusted by the hand wheel.)

just like a very fast pair of scissors – they mainly have two surfaces of metal touching at any one time (unlike a single rotary blade, which acts much more like a scythe blade that thrashes through the grass, without any other metal blade contact).

If the cutters catch something solid like a small stone, nail or twig in-between the blades it can knock the blade setting out, bend the blade or take a nick out of the metal.

On most modern British domestic cylinder lawnmowers, the cylinder moves onto a fixed bottom blade. Before you adjust any cutting cylinder, you need to check all the cutters. If these are damaged in any way it will be difficult to set the mower to cut correctly. Don a pair of gloves and inspect each cutter by turning the cutting cylinder slowly. Check the edge of each blade for any nicks, bends and for sharpness. If any of the blades has a small nick, if left it will not set properly without making a clanking or clicking noise every time it goes around (this will drive you insane if left clattering). Small nicks can easily be removed with a hand file. If any of the blades are bent, you will not be able to set the cutter at all; the blade has to be straightened to realign the cutter profile. Hold the blade steady with one hand (remember you've still got your gloves on), and tap the blade on the bend with a hammer to straighten it. If the blade is badly bent you might need to give it slightly more than a tap, or preferably take it to a professional, as it will be extremely hard to reset properly. If the actual blade edges are dulled and blunt, the only proper way is to take it to a service workshop and have it reground on a cylinder cutter grinder.

Another way of sharpening the cylinder without a grinding lathe – providing there is no damage on the blade – is a method called back lapping. This is achieved by spreading an oil and grit type grinding paste (similar to valve grinding compound for car engines) onto each blade and turning the blade backwards (the opposite direction to cutting) to grind it in. Depending on the model, remove the cutter chain or belt. Attach a brace to the nut on the cutter shaft, set the cutter to the bottom blade as if you were setting the machine to cut and turn the cutter slowly backwards by hand with the brace. Ensure all the paste is cleaned off all the metal parts when finished, as the bearings and chains will be damaged if contaminated with grinding compound.

However, there is also another easy and cheap do-it-yourself solution, albeit more temporary. It is a simple, inexpensive device with an abrasive emery type surface that clips onto the bottom blade. (Before dismantling electric machines always unplug from the electricity supply, on petrol machines disconnect the spark plug lead, and wear stout gloves when handling the blades.) Adjust the sharpening strip to the length of the bottom blade by cutting it to length. Using the mower's cutter adjustment screws, move the cutting cylinder away from the fixed blade. Clean the fixed bottom blade, then push the sharpener into place with the abrasive side facing the cutting cylinder. Tighten the adjustment screws equally on both sides, until the cylinder just touches the abrasive strip (as you would do to set the machine to cut) and so that it can still be carefully rotated by hand. Ensure that you can revolve the cutter without binding or being over tightened, so that it sharpens each blade as it turns. On hand mowers, push on a firm surface for about two or three minutes. On electric machines reconnect to the electricity supply, switch on for about one minute. On petrol machines reconnect the spark plug, start up and run for about one minute. Switch off the power and disconnect, check that all the cutting blades are sharp. If not, repeat the above procedure. Remove the sharpener and readjust the gap between the cylinder and fixed blade so that they just touch. After using for a few times, recheck and test that the blades just touch. This type of sharpening kit can be obtained from most garden machinery agents.

Atco Commodore cutter
mechanism, exploded view.

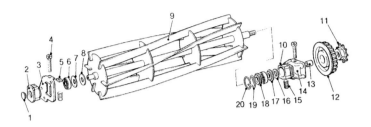

Illus. No.	Description
1	Closing plug
2	Bearing housing R.H.
3	End bracket assembly - R.H.
4	Screw for cutter adjustment
5	Nut for cylinder adjustment
6	Bearing
7	Plastic seal - R.H. bearing
8	Washer for cutter seal
9	Cutting cylinder - B17
	Cutting cylinder - B20
10	Bearing housing - L.H.
11	Cutter chainwheel
12	Pulley 46 groove
13	Spacing tube
14	End bracket assembly - L.H.
15	Cutter adjustment spring
16	Felt seal
17	Dished washer - L.H. bearing
18	Plastic seal - L.H. bearing
19	Wave washer
20	Circlip for L.H. bearing housing

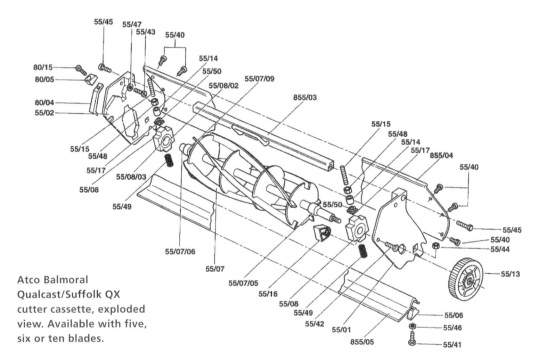

Atco Balmoral
Qualcast/Suffolk QX
cutter cassette, exploded
view. Available with five,
six or ten blades.

To set a cylinder cutter, each revolving blade has to barely touch or just 'kiss' the bottom blade all the way along its length. On the majority of most modern domestic British lawnmowers, on top of each side of the cylinder shaft, there is either an adjusting screw, nut or adjusting bolt.

1. Start by wearing gloves and making sure the blades and bearings are in good order.

2. If just a fine tune is required, adjust the screws a very small amount at a time clockwise either side of the cylinder to put the blade closer to the bottom blade. Adjust anticlockwise to move the blade away from the bottom blade.

3. Turn the cutter cylinder by hand carefully and slowly in the cutting direction. To test the cut use a length of stiff paper at several intervals along the length of the bottom blade. As

Ransomes described over 160 years ago, 'The cutter adjustment should be set to a gentleman's visiting card.' We suggest a narrow length of something similar to the thickness of the front cover of a magazine (160 gsm). (You could always test the blade setting with grass, but watch those fingers, and use long grass.)

4. The cylinder is correctly adjusted when the blades just lightly brush or kiss the bottom blade evenly across the full width of each blade of the cutter.

5. If the cutter is well off set, it is worth removing the chain or belt cover and check the cutter chain or belt tension, as when the cylinder is moved downward it also tightens the chain or belt which may then need to be readjusted.

Note: there are increasingly more machines coming onto the market with 'frictionless' cutting cylinder blades; these are set a fraction 'off' from touching the fixed bottom blade.

An extra adjustment tweak often fitted on many older models is the grass delivery plate (often called the 'throw plate'). This is situated

Suffolk and Qualcast cup and cone type cutter bearing.

behind the cutting cylinder, and it can be adjusted forwards or backwards as necessary to fine-tune and improve the angle of the grass throw into the grassbox. Slacken off the two screws or bolts on the top, or at the sides of the deflector plate. Move the grass deflector forwards or backwards evenly on both sides. Retighten and check the fixings when the ideal position is reached.

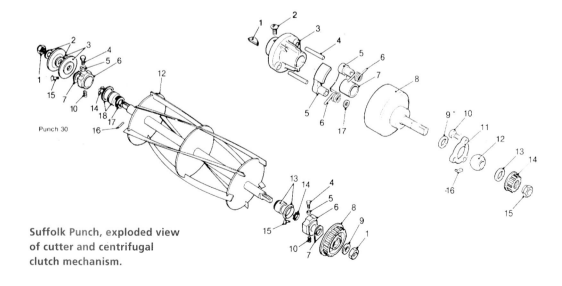

Suffolk Punch, exploded view of cutter and centrifugal clutch mechanism.

Punch 30

ABOVE: **Adjusting the grass delivery plate to fine tune the angle and improve the throw of grass into the grassbox.**

TOP RIGHT: **Testing and setting a cutter cylinder with paper. A hundred years ago the operator would be required to 'set the cutter with the thickness of a gentleman's visiting card'.**

BOTTOM RIGHT: **A cutting cylinder being reground on a professional grinding machine, ensuring every blade is equidistant to the centre of the cutter shaft.**

On most cylinder lawnmowers the blades can be re-sharpened repeatedly until the blades are almost worn down to the 'spiders' (the blade holders, which attach onto the central cutter shaft).

Cylinder lawnmowers are designed to cut lawns with short grass, but slightly longer grass can be tackled by fitting two narrow front rollers with a slightly larger diameter, in place of the standard full front roller(s). Fit the two narrow rollers at each side of the chassis separating them apart with a wide spacer. (A piece of garden hose pipe can often make an admirable spacer.) Front roller kits are also available for some models. This easy-to-fit modification allows longer grass to stay standing up and allows the grass to go straight into the cutter without the roller bending the grass down. The extra wide diameter of the roller allows the fixed bottom blade to be raised higher off the ground. To calculate the maximum height of grass that a cylinder machine will cut without overloading:

1. Place the machine on a flat surface – a garden flag is ideal.
2. Adjust the cutter to its highest position by moving the roller(s) downward; this raises the bottom blade upward (this sounds a contradiction but is correct).
3. Measure the distance between the ground and the top of the bottom blade.
4. Measure the distance between two of the cylinder blades.
5. Then add the two measurements together to calculate the maximum height of the grass your machine will cut without struggling.

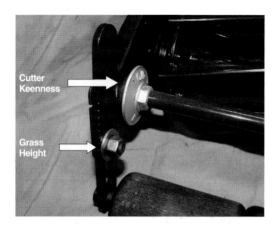

Webb Whippet, Wasp and Witch: height of cut and keenness of cut adjusting methods.

The cutter height can be adjusted via a hand wheel on the front roller (Qualcast Punch).

The height of cut on most cylinder mowers can be adjusted independently by roller brackets fixed to the front at each side of the machine (some electric and side-wheel cylinder models are fitted at the back). They often have elongated holes, which can be locked by a hand wheel or nut and bolt. Other adjustments may be via a single knob or wing nut, depending on the model. Whatever the model, it's the rollers (or roller) that are moved up or down to alter the cut height.

sprockets, shafts and the chain. Note: when adjusting a cutter cylinder downwards it also alters the chain tension; check that the chain has not been over-tensioned too tightly.

To maintain a chain, especially if it has had a lack of lubricant and is seized, remove the chain and soak it in paraffin. Ensure all the links are moving freely before refitting. Check the wear in the chain by laying it out on a flat surface; push and pull the chain to see if there is any loose play between the links. Note, if the chain is worn and

The drive chain

There are only a few different sizes of chain fitted to common lawnmowers (none of which are the same size as bicycle chain). Virtually all chains on mowers can be adjusted. To achieve the correct adjustment, rotate the chain slowly around on the sprockets, find the tightest spot and adjust the tension at that point, with approximately 1cm play on the longest side. Ensure the chain does not 'bind' or is too slack when turning. If the chain is adjusted too tight or left binding this can cause premature wear on the bearings,

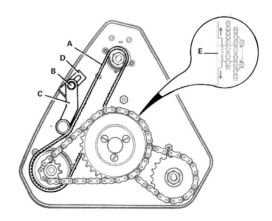

Atco Commodore chain drive, showing roller drive clutch disc.

stretched and cannot be tensioned enough by the adjuster, do not be tempted to shorten it by a link or a half link. In these circumstances the chain is worn out and it would be better to replace the chain. If you remove a link, however worn or stretched the chain is, it will be too short. If the chain is worn, it will cause the sprockets to wear; if the sprockets are worn the chain will wear prematurely.

Most chains are joined together by a split link.

To remove a split link, use pointed pliers and push the open end of the split link clip toward its link pin. When refitting a split link ensure the rounded end of the link is facing forward in the direction of travel – this avoids the possibility of the link becoming unclipped when the chain is turning rapidly. If refitting a chain without removing the split link, ensure the split link is on the outside face, so you can check the split link is facing in the right direction.

Chain, split link and clip, showing the correct direction of the split link.

Twin sprocket and chain, showing the direction of the chain split links.

Atco Commodore transmission showing chain tension adjuster and split links.

Exploded view of Atco engine/cutter with Ferodo clutch disc assembly: two Ferodo discs transmit the power from the engine to the cutter. Heavy-duty models use three larger discs. The engine can be engaged and disengaged, just like a car.

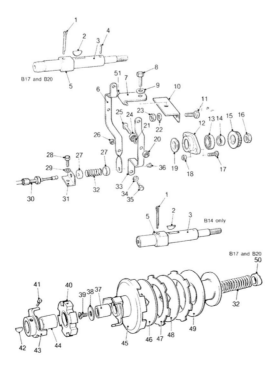

Illus. No.	Description
1	Split pin
2	Key for shaft
3	Clutch shaft - B14
	Clutch shaft - B17
	Clutch shaft - B20
4	Roll pin - B17 & B20
5	Driving square
6	Clutch fork upper
7	Clutch fulcrum assembly - B14
	Clutch fulcrum assembly - B17
	Clutch fulcrum assembly - B20
8	Set-pin for fulcrum to bracket
9	Plain washer
10	Clutch fulcrum assembly
11	Screw for clutch fulcrum
12	Bearing housing
13	Ball race
14	Spacer for pulley
15	Pulley 22 groove
16	Nut for clutch pulley
17	Socket cap screw
18	Self-locking nut
19	Thrust collar - B14
20	Nut
21	Plain washer
22	Washer for clutch fulcrum
23	Nut for clutch fulcrum
24	Bearing
25	Shoulder bolt
26	Nut for clutch fork clamping
27	Spring cup
28	Screw for bracket fixing
29	Washer for bracket fixing
30	Cable assembly
31	Clutch cable bracket
32	Clutch spring
33	Clutch fork lower
34	Washer for clutch fork
35	Screw for clutch fork
36	Spigot washer
37	Bush for boss
38	Washer for bearing retainer - B14
39	Screw for clutch shaft - B14
40	Rubber coupling
41	Screw for engine coupling
42	Key
43	Engine coupling assembly - B14
	Engine coupling assembly - B17 & B20
44	Bush for coupling - B17 & B20
45	Driving member assembly
46	Driving plate - thick
47	Friction plate - Ferodo
48	Driving plate - thin
49	Withdrawal sleeve assembly
50	Adjusting collar - B17 & B20
51	Pin for fulcrum

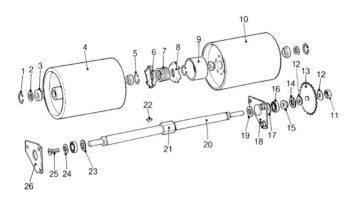

Some typical exploded views of the working parts of a common Atco petrol domestic lawnmower (Commodore), offering an explanation of how many parts are involved and a better understanding of how they are assembled and work together. Many mower clutch mechanisms are similar to the workings of a car, but to a smaller scale. ABOVE: Split rear roller assembly showing differential gear, designed for easier turning. BELOW: Chassis assembly.

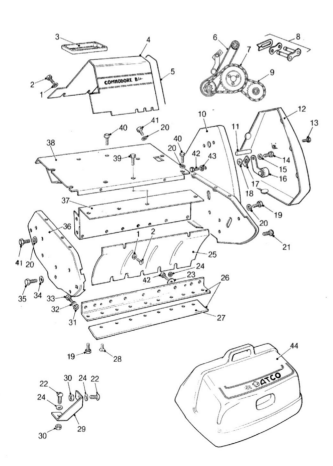

CHAPTER FIVE

ROTARY MOWERS

There are three types of rotary mower: four-wheel, rear-roller and mulch; and four types of rotary chassis construction: aluminium, steel, stainless steel, and various types of plastic. Steel chassis machines are durable and inexpensive to manufacture. Many economy mowers, especially in rotary, are made of steel; the advantage is that they are cheap to produce, but the disadvantage is that they will rust. Aluminium chassis are more costly, but are stronger and will last much longer, and they have the advantage of not rusting. Plastic chassis vary enormously from the cheapest of machines to some of the most expensive. The advantages of plastic, especially in electric rotary machines, are that they are extremely light-weight, and very low cost. Some plastic chassis mowers are extremely strong (stronger than steel in a lot of cases) and they can also be a heavier weight. All plastic machines have the advantage of non-rusting and the quality of the machine is normally reflected in the price. The fourth option, stainless steel, is durable, lightweight and will not rust, although it can be a tad more costly. The disadvantage at present is that there are only a few manufacturers using stainless steel and the choice of models is limited.

Rotary blade sharpening

Rotary blades require regrinding and balancing every season or every twenty-five hours' running time, or checking more often, especially if the blade has come into contact with anything more solid than grass. A damaged, out-of-balance blade can drastically shorten the life of the mower and the engine if left unchecked. (Would you carry on driving your car if the steering wheel started shaking? No – you would get the wheels balanced.) Unbalanced blades will cause seemingly unconnected problems, such as handles breaking or snapping, nuts and bolts coming loose, wheel and axle wear, various noises, and many more. When a blade is damaged, especially on basic rotary machines (this normally happens after hitting a solid object like a tree root, a hidden stone, a concrete kerb, fallen branch, a child's toy or, if the cut is too low, just digging into the soil), if it comes to a sudden stop, something has to give. The average speed of a rotary blade is over 160 mph; it spins at approximately 3,000 rpm. The main damage is concentrated on the blade, the blade mounting or the engine crankshaft. Depending on the force of the impact, if a mild impact is sustained, the blade, blade mounting and crankshaft can be checked, and the blade can be reground and balanced; if a heavier impact is sustained (and you only need the one) and a vibration is detected by holding the handle grips, further inspection is required and it is best to let a professional deal with it. In severe impacts the main engine crankshaft can bend – when this happens it is often terminal for the engine. (If this happens, it is worth checking your household insurance policy as you may be covered for accidental damage.) There are several machines on the market with impact blade

and crankshaft protection. These types can be very worthwhile considering for your next purchase, especially if you are prone to cutting anything other than grass. Some rotaries which have blade and crankshaft protection consist of several different types of mechanisms, including a moveable swing tip blade. This simple type of blade

18in Hayterette four-swing blade system. The swing blade system allows the blade to take an impact without damage to the engine, by the blade swinging out of the way. The blade automatically swings back to its normal position by centrifugal force.

ABOVE: The 18in, 20in, and 22in Rovers have a two- or four-swing blade system, protecting the engine crankshaft on impact with an object like a stone or tree root.

MIDDLE LEFT: Hayter Harrier blade clutch disc, fitted on Harrier petrol Spirit and Ranger models. This system has a conventional single blade fixed onto a metal disc, which acts as a slipping clutch if the blade hits an object. It is also covered with a lifetime crankshaft warranty if damage to the engine occurs after an impact.

BOTTOM LEFT: Honda 17in, 19in and 21in Rotostop system. This blade system is fitted on some Honda and Hayter mowers (Blade Brake Clutch: BBC). The clutch acts in a similar way to a car disc clutch, allowing the blade to slip if impacted. There is an added safety and convenience feature: when the operator lets go of the operator presence control, the blades automatically stop, but the engine carries on running. This also has the benefit of not having to restart the engine each time you let go of the machine.

system is made up of two or four moving blades fixed by a single bolt onto a central disc. When the blade impacts an object it swings out of the way. Another system is a solid blade bar centrally fitted over a disc; when the blade has an impact the whole blade moves on the disc acting like a clutch. Another system is a blade brake clutch mechanism, which separates the blade from the engine via a clutch disc mechanism, operated via a clutch cable and lever. This system also has the advantage of being able to stop the blade without stopping the engine.

To remove a rotary cutter blade:

1. Drain the fuel by running the engine till the tank is empty and the engine stops.
2. Remove the spark plug cap and let the engine cool (this ensures the engine will not start up if the blade and crankshaft are turned).
3. If fitted, depress the blade brake lever. Then pull the recoil slowly till the rope goes stiff. This closes the engine valves and helps prevent the oil from entering the cylinder head.
4. Release the blade brake lever (if fitted).
5. Ideally tip the machine with the spark plug facing upwards – this position will ensure that the oil will not contaminate the cylinder head carburettor, air filter and silencer. Sometimes this position may not always be possible, depending on the model, as there may be parts of the mower restricting access to the blade. In which case, remove the air filter to ensure that it won't be contaminated by oil, and turn the mower on its side. Consult your handbook or your authorized service dealer as to which side it can be tipped. If the oil and fuel are removed the engine can be tipped either way.
6. Most rotary blades are fitted with one single bolt in the centre, or two bolts mounted onto a flange. In most cases single-blade machines have a standard anticlockwise thread to

remove the blade bolt. Ensure you have the correct size spanner, use a good quality socket set or ring spanner. (An adjustable spanner will not do: the bolt will be tight, especially if the blade has not been removed for a while, and rounding off the bolt flats will make it all the harder to remove.)

7. Put gloves on and hold the blade firmly with one hand whilst turning the spanner with the other, or chock the blade with a piece of wood to assist.

Note: to avoid an accident, do not work on a cutter blade unless you have disconnected the spark plug lead. Blades are sharp, so always wear gloves. Do not rotate tools toward the cutting edges – this avoids injury if the tool should slip. Ask for assistance when turning the mower on its side. Use genuine replacement spares especially when blades are involved.

Once you have removed the blade, check for any damage. Is the blade bent? Are the blade tips damaged? Is the blade worn or does it show any signs of cracks? If anything is suspect at all, replace the blade. Some rotary machines are fitted with two- or four-type swing blades bolted to a disc; provided the disc is not damaged or bent it is not necessary to remove the centre disc bolt on this system. On these models its easier and quicker just to replace the blades. It is also worth replacing the bolts as well when changing these blades in case of any wear or fractures. Regularly check the central blade-securing bolt for tightness.

Providing that the ends of the cutter bar are not worn down, the blade tips can be sharpened with a file. Depending on the cutter length the blade tips are between 1 inch (25mm) and 3 inches (75mm) in length. The actual grasscutting is only done on the last few inches of the blade, so it is not necessary to sharpen the whole length of the blade. Copy the angle of the blade tips, which is approximately 30°–45°.

Balancing

After the blade has been sharpened, the blade must be balanced. The balancing of the cutter bar can be more important than the sharpness. On a rotary cutter blade, both blade edges must be sharpened equally to ensure its balance. Clean the blade off thoroughly with a stiff wire brush as quite an amount of thick, hard, grass debris build-up can accumulate on the blade, causing an imbalance and aerodynamic vibration. The cutter blade tips can be filed with a flat file at the ends of the cutter bar to restore the cutting performance.

To balance a cutter bar, put a screwdriver through the centre hole of the cutter bar. Hold the cutter bar and screwdriver in a horizontal position. Release the blade and let it find its balance position; if the blade does not stay level, file more off the end that tips downward until the blade balances horizontally.

To refit the cutter blade, ensure the turned-up edges are facing towards the engine. These turned-up edges, commonly known as wing tips, are shaped to create an air draught to throw the grass into the collection box. Check for any wear or damage on the centre hole, and refit all washers, spacers and bolts carefully in the order in which they were removed. Also check if any of the fittings or mountings are cracked or damaged – if so replace them, as it's not worth risking any suspect part of a cutting system.

Alternatively, take the blade or the mower to your local garden machine service dealer. All good service workshops will be equipped with a professional blade balancer. The blade balancer also checks if the blade is true and not bent or distorted. Ask to have your blade reground and balanced. This can be done inexpensively and often quickly (some service dealers may offer a while-you-wait regrind and balance service, especially if it is not in the height of the season). The best time to take your mower for an annual service is any time when the grass is not growing; remember, 11°C is the magic temperature.

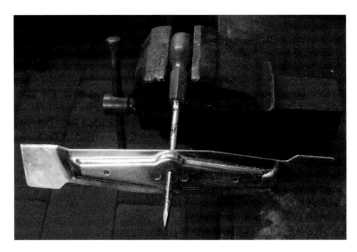

ABOVE: A simple way of balancing a rotary cutter blade 'DIY-style' at home.

RIGHT: Professional blade balancer; note the adjustable gauge to check if the blade is true and not bent or distorted.

LOOKING AFTER CABLES AND WIRING

When folding the handles for storing or transporting, before fully folding the handles check that the control cables are not going to be trapped or kinked at the pivot bolt of the handle. Also, don't over-tighten the handle pivot bolts – they can be stronger than the tubular handle, which can easily distort the tubular steel leaving the handle with a poor fit that cannot be tightened.

After purchasing a new mower that has been supplied in a box, check when assembling the handle that the cables and any wiring are not going to be snagged when the handle is folded; also check the cables are not twisted and are free to operate. If assembled twisted, the cable will be stiff and may not operate smoothly or will just create premature wear.

If you don't do any of the service work in this book, but you watch out for the cables when folding the handles, it will save you at least the cost of this book and possibly a big chunk out of the cost of your repair.

ABOVE: **Incorrect cable position with the consequence of the cables being kinked.**

RIGHT: **Correct position of control cables on top of the handles before folding.**

CHAPTER SIX

BRIGGS & STRATTON ENGINES

A handy little device from Briggs & Stratton is the electrical service indicator, which will fit to any petrol mower. It's extremely easy to fit onto the handlebar and tells you when service is required, just like on many modern cars. Called 'servisexpert', it indicates: (1) when to check the oil level; (2) when to change the oil; (3) when to change the oil and air filter; (4) when to change the oil, air filter and spark plug; and (5) when a full service is due.

Note: always remove the spark plug cap when working on petrol mowers. This prevents the engine from firing and accidentally starting.

Serial numbers

Briggs & Stratton engines have about a twenty-figure unique model number designation system. The model numbers are amongst the easiest to find – all the numbers should be stamped on the engine cowling, normally at the top or on the side of the engine, above or next to the spark plug or above the carburettor. The numbers denote the model, type and code of the engine. The first one or two numbers indicate the approximate size of the engine in cubic inch displacement (5–61). The next number indicates the basic engine design series (0–9 or A–Z), and relates to cylinder construction, ignition type and general configuration. The next digit indicates the orientation of the crankshaft (0–4 or A–G = horizontal; 5–9 or H–Z = vertical). The

digit after that (0–9) indicates the type of bearings and whether the engine is fitted with a reduction gear or auxiliary drive. The final digit indicates the type of engine starter (0–9 or A). If you are unable to find a serial number stamped on a Briggs & Stratton engine it is very possible that the engine cowling may have been changed at some time.

Starting the engine or motor

One of the most common faults on petrol and electric mowers is that the engine or motor will

Servisexpert: an engine maintenance monitor, which can be fitted to most petrol engines and set up in seconds.

UNDERSTANDING BRIGGS & STRATTON SERIAL NUMBERS

To Identify a typical Briggs & Stratton engine, model 253447, type1234-01, code 01082301:

Model 253447

25 = 25 cubic inches (engine cylinder bore size)

3 = engine design series

4 = horizontal crankshaft

4 = ball bearing, flange mounting, pressure lubrication

7 = electric start (12 or 24 volt), gear drive with alternator

Type 1234-01

This number identifies the mechanical parts, colour of the paint, decal design, governed speed, and the original equipment manufacturer.

Code 01082301

This number identifies the date of manufacture.

01 = year (2001)

08 = month (August)

23 = day (23rd)

01 = assembly line, manufacturing plant

thing wrong. (You would be amazed how often a mower is taken into a repair shop for a no-start problem when all it required was some petrol in the tank.)

Have a good look in the tank: you may be able to see fuel in it, but there may not be enough fuel to fill the carburettor; some tanks do not drain dry, especially on an engine were the tank is situated below the carburettor, common in domestic Briggs & Stratton engines. On these models the fuel is sucked upwards and it does not drain the tank completely of fuel, unlike most gravity-fed models. On these engines ensure that the fuel tank is at least half full, especially for a cold start. If the engine is fitted with a primer bulb (some are red, some are black), does the primer bulb feel as though it is squirting fuel when pumped, and does it return to shape after depressing? On many common Briggs & Stratton engines such as Classic, Sprint and Quattro engines, to see and check if there is fuel squirting, remove the single screw on top of the air filter. Remove the complete filter assembly. Be careful when removing the filter not to get any debris dropping inside the carburettor. Push the primer bulb – you should see a fine jet of fuel squirt into the engine. If this is not the case, further inspection is required. Whilst the filter is off, check if it is clogged with grass.

Cleaning or replacing the filter

not start. Firstly we will look at the petrol engine. You've pulled the cord several times and not got a murmur from the engine. You can do several things: carry on pulling the recoil until either you break the rope or the recoil, or just lose patience. If you need more than five or six pulls to start an engine, do some very simple quick checks first before you carry on pulling, as there is either a fault with the engine or you may be doing some-

There are two main types of filter: sponge type and paper type. If you have an engine with a sponge filter, now's the time to put some disposable gloves on, unless you just want to buy a new filter. (Note: if you are not sure of your engine model, take the filter to the service dealer for the correct match.)

SCREW

ASSEMBLE ONE OF THESE LOW POINTS TOWARDS NARROW EDGE OF ELEMENT

CUP

ELEMENT

BODY

ASSEMBLE ELEMENT SO LIP EXTENDS OVER EDGE OF AIR CLEANER BODY

LIP WILL FORM PROTECTIVE SEAL WHEN COVER IS ASSEMBLED

A

B

ENGINE OIL

C CLOTH

C

LEFT: Method of foam type filter removal and cleaning. Now is the time to put the rubber gloves on.

Good clean paper filter element.

Half-clogged paper filter element. The filter can be removed, and providing the paper does not looked clogged, debris can be tapped off and the filter re-used.

Completely clogged paper filter. Throw the filter in the bin and replace it.

Normally check the air filter every twenty-five hours (generally considered as one year's work), more often if you've been cutting in dry and dusty conditions. If it takes six hours every week to cut your lawn you need to check the filter every four weeks. (If this is the case we suggest you buy a bigger mower, probably a ride-on.)

1. Remove screw or wing nut.
2. Remove air cleaner carefully to prevent dirt from entering carburettor.
3. Take air filter apart and clean.
4. Wash foam element in paraffin or liquid detergent and water to remove dirt.
5. Wrap foam in a cloth and squeeze dry.
6. Saturate foam with clean engine oil. Squeeze to remove any excess oil.
7. Reassemble parts, refasten to the carburettor with the filter screw or wing nut.

Briggs & Stratton tractor paper filter element, 10hp upwards. Foam filter: 2–8hp engines.

The oil residue left in a foam filter is there to catch fine particles of dust. Wipe and clean the filter casings then refit the filter back in the housing. Alternatively, just replace the filter with a new one especially if the sponge shows any signs of deterioration.

The paper filter type, if not too clogged, can be tapped to dislodge any surface debris. If more than tapping is required it is best replaced with a new paper element. Most filters are white in colour; if it has turned grey with dust particles, replace it. If any part of the filter has been contaminated with oil or fuel replace it, or if there is a hole or damage in the filter. The air filter is the main defence to keep the internal engine parts clean; once breached, the carburettor will have to be removed and cleaned. You can also cause air filter contamination by tipping the machine. These filters are designed to let just the correct amount of air through but stop damaging microparticles entering the engine, which will shorten the engine's life. A blocked filter causes the wrong air and fuel mixture, which chokes the engine making it run erratically, blow smoke out, smell, lose power, and the engine will use more fuel. On some modern mower engines the fuel and air mixture jets have been factory preset and cannot be adjusted without professional tools, and on some engines the adjustment screws have been removed altogether. This makes it all the more important to have a clean filter. Unlike on some engines – especially older ones which have several fuel mixture and air mixture screws to adjust for smooth running – they can compensate for any discrepancies in the condition of the fuel, filter, carbon build-up, or spark plug. It is often worth letting your local service dealer finetune an engine to get it right: if you have a finicky engine it is not worth spending a long time adjusting the mixture, and it is better to leave this job to a professional.

Briggs & Stratton carburettor with the filter removed, showing the primer bulb and governor linkage.

When changing a paper filter, make a note of the position of the filter when it came out and ensure you put the filter back the same way. If you are going to tip a mower on its side with a paper filter fitted, it's worthwhile removing the filter temporarily to avoid oil contamination spilling onto the filter (this will destroy the working properties of the filter and render it useless). Ensure debris does not enter the engine whilst the filter is removed; protect the carb opening by covering it with cling film.

MAINTENANCE LIST

Every time:

- Change the oil after five or six hours of use from brand new.
- Check the oil before starting the engine.
- Remove any debris from around the engine and blades.

Every 25 hours:

- Change the oil, especially if used under heavy load or in hot conditions.
- Replace the spark plug.
- Clean or replace the air filter.
- Check the fuel pipe and tank for any debris.

- Check the recoil rope for wear.
- Check the engine cooling fins and engine cowling are clear of debris build-up, especially if you have been scarifying.
- Check around the carburettor springs and linkage for debris build-up.
- Check all visible nuts and bolts for tightness.
- Re-sharpen and balance rotary blades.
- Regrind and set cylinder blades.

Note: if the machine has been running in dusty, dry or dirty conditions, check more frequently.

CHAPTER SEVEN

ATCO, QUALCAST AND SUFFOLK PETROL LAWNMOWERS

For that perfect traditional formal lawn, a rear roller is required and the heavier the better. It is the roller that forces the grass in one direction and then in the opposite direction creating stripes. The heavier the mower, the more the stripe will be defined. Cylinder lawnmowers accentuate the striping because the grass is always cut in one direction only, unlike a rotary. The simplest way for the best stripe effect is to cut around the perimeter first to provide a turning area. Then cut in straight lines either up and down your lawn or side to side, with each cut slightly overlapping the previous one. Other patterns can also look effective, such as diagonal stripes or a chequerboard effect, cutting in one direction and then at right angles. If the area is large enough, circles and diagonals can be formed, like at some top football grounds.

you require (except for 12in electric QX models, which have only a five-blade cutter and a scarifier cassette option available at present). It's worth considering having more than one cutter cassette, giving you several easy options as a spare, which can be changed in just a few minutes and

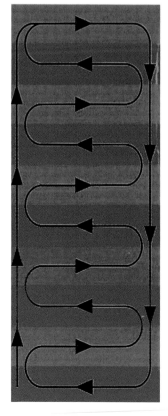

QX cassettes

The Atco Balmoral and Windsor, and the Qualcast Classic and Suffolk Punch petrol and electric QX (Quick Exchange) lawnmowers have interchangeable cassettes on 12in, 14in, 17in and 20in models, making them versatile two or three machines in one. As well as scarifier cassettes, five-blade, six-blade and ten-blade cutter cassettes are also available, depending on the finish

Mowing guide, showing the direction of cut for a traditional striped finish with a petrol or cordless lawnmower. Cut around the edge first to allow for turning.

it can in some cases come in very handy. If you want the grass to have a finer finish, perhaps in mid-season when the grass is lush, drop the ten-blade cutter cassette in. Alternatively, if the grass is a bit on the longish side when the grass has been left for a while, drop the five-blade cutter cassette in, as it will tackle it a lot better. Also if you catch an object in the cutter, within a few minutes you're back up and running again, without taking a trip to the service department where you may have to wait for a repair and be without a mower, especially in the height of the season.

To change any of the above models from a lawnmower to a scarifier, or vice versa:

1. Wearing gardening gloves, remove the grass-box cradle. On Atco Balmorals there is a spring-loaded clip with a ring attached on the right-hand side of the cradle, pull the ring sideways. On Qualcast Classic and Suffolk Punch models, gently squeeze the cradle bars approximately ¾ inch (20mm) towards each other.
2. Remove the four hexagon set screws fitted on the side of the plastic drive belt cover with a 5mm hexagon key normally supplied with the machine and also with each scarifier and cutter cassette.
3. Remove the two retaining hexagon screws on each side of the cassette; these are fitted above the cutter shaft on each side.
4. Grip the centre handle of the cassette firmly and pull it upward and forward; the cassette will come out. Sometimes a good tug may be required if the cassette hasn't been removed for a while so obtain assistance if it seems a tad stiff. The cassette will come out complete with the toothed drive gear.
5. Reverse the procedure to fit any of the cassettes.

On early electric QX models only (remember, there's always an exception on lawnmowers) a belt pulley is fitted instead of a toothed gear. Before removing the cassette, remove the belt off the cutter pulley (still with your gloves on), then remove the cutter pulley, by first chocking the blades from turning with a piece of wood (a hammer shaft is ideal). Grip the pulley with a gloved hand; it has a left-hand thread so turn it clockwise to remove the gear. The pulley is then fitted onto the cassette accessory. A quicker and easier method (as sometimes these pulleys can be tight to remove) is to purchase an extra pulley and

ABOVE: **Atco Balmoral, Suffolk and Qualcast Punch transmission.**

BELOW: **Atco with the transmission pulley removed, showing hidden drive gears.**

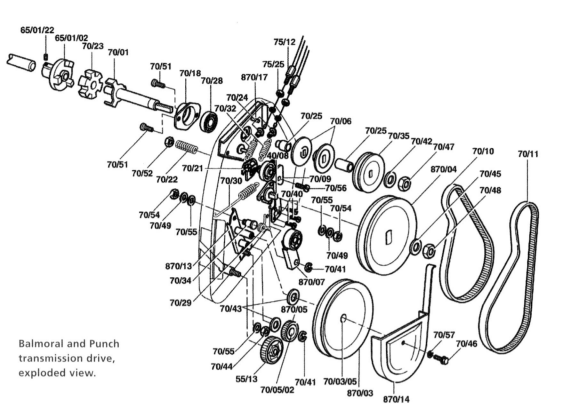

Balmoral and Punch
transmission drive,
exploded view.

adaptor, which comes as an assembly, and leave it fitted permanently on your cassette accessory. The part number for the pulley kit is F016L02161 and would be available at all garden machinery agents. Adaptors are included with the cassettes on later models.

On petrol Atco Balmorals, Suffolk and Qualcast Punches there is no need to remove the plastic toothed gear on the cutter shaft; whilst the cassette is removed the gear can be checked for wear. The gear is fitted with a left-hand thread – if it needs to be replaced turn the gear clockwise to remove. If the gear is tight, place a piece of wood (e.g. a hammer shaft) inside the blades to lock the cutter. Put gloves on and turn the gear clockwise.

Whilst the cutter cassette is out it is a good idea to look at the cylinder blades, checking for any bent blades or blade chips. Check the bottom fixed blade for straightness or chips (look down the length of the blade), and check that it is not worn; the cutting edge should be straight and not chamfered off. With gloves on, check the cutter bearings; there are a couple of very quick tests:

1. Grip the central shaft of the cutter firmly and pull the shaft up and down. If you notice any up and down movement it is likely that the bearings are worn and require replacement.

2. Check that the cylinder cutter is not binding on the bottom blade; if it is, slacken off the cutter adjusting screws anticlockwise a few degrees. Spin the cylinder slowly (make sure your fingers are out of the way). The cylinder should spin freely and smoothly.

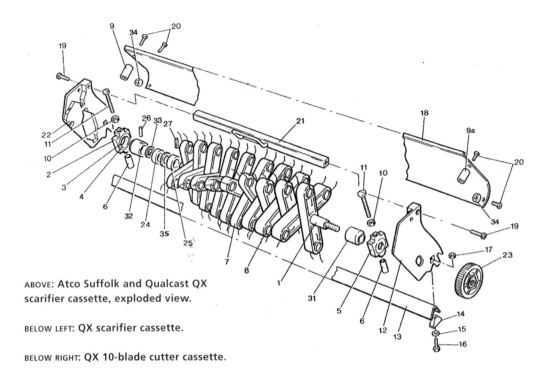

ABOVE: **Atco Suffolk and Qualcast QX scarifier cassette, exploded view.**

BELOW LEFT: **QX scarifier cassette.**

BELOW RIGHT: **QX 10-blade cutter cassette.**

Suffolk engines

The Suffolk engine (made by the Suffolk Iron Foundry) was made in several guises. The most popular engines were 1½hp (75cc) and 2hp (98cc and 114cc), fitted to virtually every domestic Atco, Qualcast and Suffolk cylinder lawnmowers from the 1960s to the 1990s. (Early Atcos were fitted with Villiers two- and four-stroke engines; later models had Tecumseh 3hp and Mitsubishi engines. Currently Kawasaki 100cc engines are fitted.) There are two main types of Suffolk engines – a cast-iron and an

aluminium crankcase. If you are not sure what type you have, to identify the cast-iron type, this has a dipstick 6 inches long; the aluminium engine block has no dipstick, just a screw plug.

The 1½hp cast-iron engine can also be identified by its six cylinder head bolts. The 98cc and 114cc 2hp models can be identified by having eight cylinder head bolts.

Suffolk engines were fitted with three main types of magnetos:

1. Early cast-iron engines: coil with separate condenser and points.

Suffolk-Qualcast aluminium four-stroke 98cc engine, exploded view.

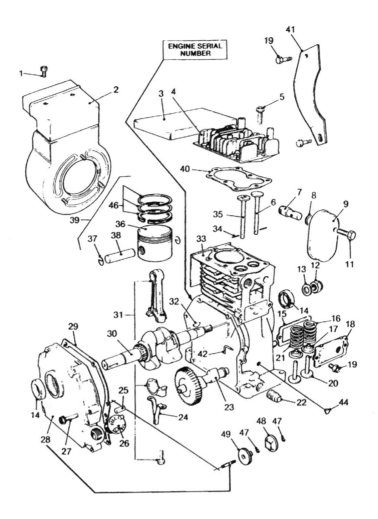

2. Aluminium engines: coil with integral condenser and points.
3. Electronic coil, no points.

To identify which type of magneto, count the number of fins on the flywheel. Early models have twelve fins; later models have twenty-four and thirty-two fins.

Zenith carburettors

Two main makes of carburettor were fitted on the Suffolk engine, the Zenith (embossed on the side of the carburettor bowl) was fitted on early models; Delorto and Tilotson carburettors on later Tecumseh engines.

Some tips on the Zenith carburettor:

• If the carburettor leaks fuel, check the float needle and seat. Remove the carburettor bowl by unscrewing the two carburettor set screws on top of the carburettor bowl.
• Remove the float and inspect both the float and the needle, the point of which should be undamaged and always face upward. If the needlepoint shows any sign of wear or damage replace it.

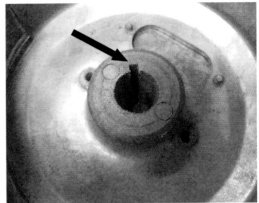

ABOVE: Three Suffolk Iron Foundry (SIF) flywheels. Twelve-fin early model magnetos were fitted with points, coil and condenser. Twenty-four-fin models have points and coil with integral condenser. Thirty-two-fin models have electronic ignition with no points.

BELOW: Suffolk magneto showing moving contact point set.

ABOVE: Inside view of a Suffolk flywheel. Check around the keyway for any signs of fracture.

- If the needle is worn and the carburettor still leaks when the float has been changed, replace the needle seat.
- Another common component to check, especially if the engine has poor starting or runs

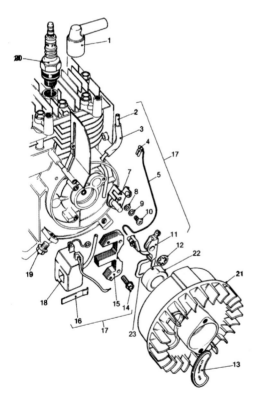

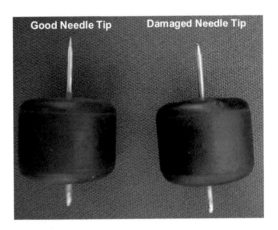

ABOVE: Zenith float and needles, in good and damaged condition. When refitting a Zenith float, to avoid damage to the float needle check that the needle point is located in the needle seat before screwing on the carburettor barrel. Check the needle-locating hole in the base of the carburettor bowl, ensuring it is clear and not clogged with debris or stale fuel varnish.

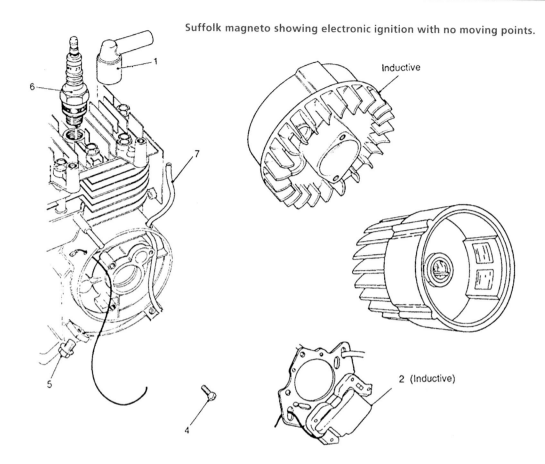

Suffolk magneto showing electronic ignition with no moving points.

rough, is the fuel mixture screw, located on the side of the carburettor. Unscrew anticlockwise and check for any wear or damage. Take a look down the mixture hole and spray with a carburettor cleaner. (If the end of the screw is broken off inside the carburettor, it is time to see a professional.) Important: when refitting the mixture screw, do not over-tighten. Turn the mixture screw gently clockwise by hand till the screw lightly touches the seat, then turn back one turn; the mixture can be fine tuned from this point.

To fine-tune the Zenith carburettor:

• Run the engine approximately at half revs to normal operating temperature.

• Turn the fuel mixture screw clockwise till the engine starts to slow, then turn the mixture screw in the opposite direction till the engine starts to slow. Then turn the screw back to the midpoint of the two positions.

• Next, run the engine at idle speed, and then turn the air mixture screw (located at the top of the carburettor) slowly till the engine starts to slow, then turn back slowly (approximately half a turn) till the engine idles smoothly.

• Check inside the bowl and remove any gunge or debris with a carburettor cleaner, especially in the bowl base where the float needle is located.

When the Suffolk Iron Foundry closed, the Tecumseh engine was fitted (named after the

Native American Chief Tecumseh). Tecumseh engines were manufactured in the USA and Italy and at one time were part of the huge Fiat car company. The latest Atco, Qualcast and Suffolk domestic models are fitted with 3hp 100cc Kawasaki engines. Atco Heavy Duty and Professional lawnmowers are currently fitted with 6hp Briggs & Stratton engines, many with key starts and trailed ride-on seat which has a unique cantilever on the seat roller to allow a tighter turning circle.

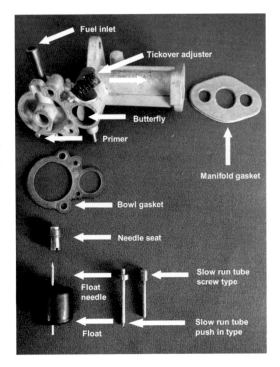

TOP RIGHT: **Zenith carburettor barrel, showing float, needle, seat and butterfly throttle.**

BOTTOM RIGHT: **Suffolk/Qualcast carburettor: exploded view of Delorto carburettor.**

BELOW: **Zenith carburettor, showing choke lever, primer, air, fuel and idle adjustment screws.**

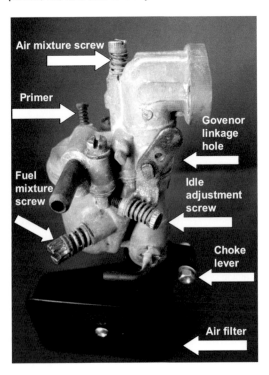

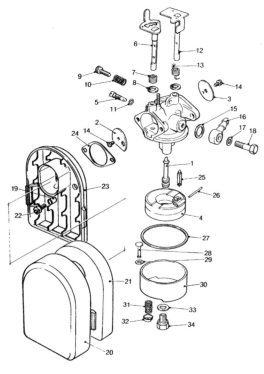

Atco Commodore, Deluxe, Mk5 and Mk7 engine clutch, exploded view.

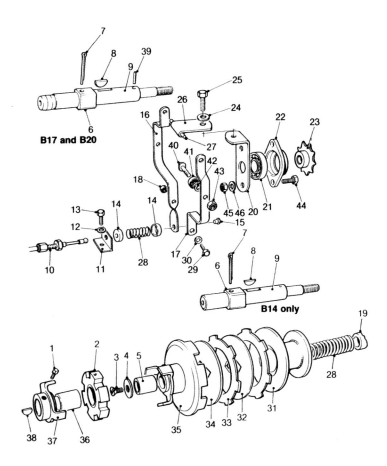

B17 and B20

B14 only

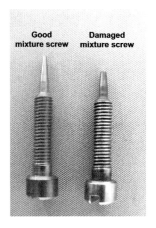

Good mixture screw   Damaged mixture screw

Good and damaged Zenith mixture screws.

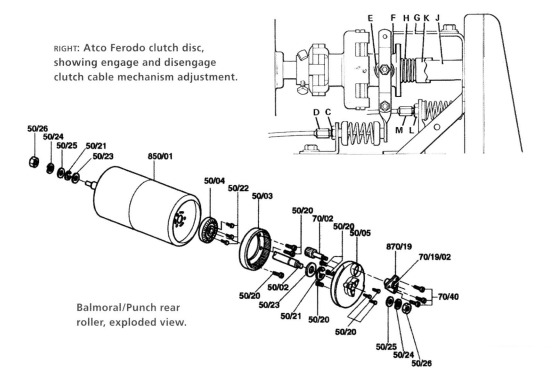

RIGHT: Atco Ferodo clutch disc, showing engage and disengage clutch cable mechanism adjustment.

Balmoral/Punch rear roller, exploded view.

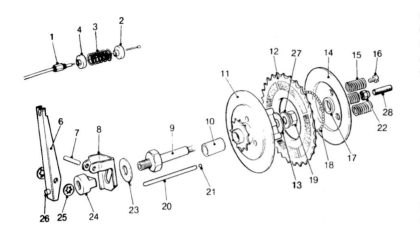

LEFT: **Atco Commodore Deluxe, roller drive traction clutch, exploded view.**

BELOW: **Atco main chassis, exploded view.**

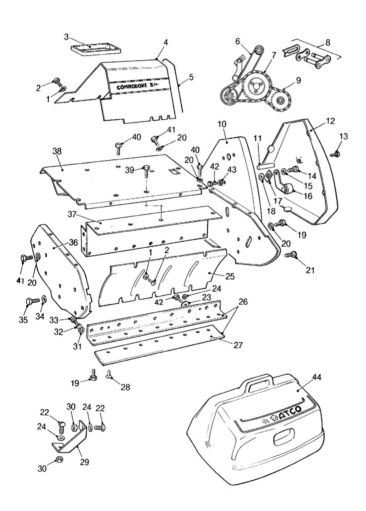

Illus. No.	Description
1	Washer for clutch cover & del. plate top
2	Screw for clutch cover & del. plate top
3	Spacing block
4	Clutch cover - B14
	Clutch cover - B17
	Clutch cover - B20
5	Clutch cover edging strip
6	Toothed belt
7	Chain
8	Connecting link
9	Chain
10	Sideplate & inner chaincase sub-assy. L.H.
11	Chain cover pillar
12	Chain cover
13	Screw for chain cover
14	Screw
15	Spring washer
16	Belt adjuster
17	Adjuster stay
18	Spacer
19	Screw for bottom blade carrier to sides
20	Washer for platform & strengthener
21	Socket cap screw for end bracket - L.H.
22	Screw for support bracket
23	Washer for delivery plate
24	Washer for delivery plate & support bracket
25	Delivery plate - B14
	Delivery plate - B17
	Delivery plate - B20
26	Bottom blade & carrier sub-assembly - B14
	Bottom blade & carrier sub-assembly - B17
	Bottom blade & carrier sub-assembly - B20
27	Bottom blade - B14
	Bottom blade - B17
	Bottom blade - B20
28	Screw for bottom blade
29	Support bracket
30	Nut for support bracket
31	Washer for bottom blade carrier fixing - rear
32	Washer for carrier fixing screws - rear
33	Screw for bottom blade carrier fixing - rear
34	Washer
35	Screw for rear roller bearing housing & end brackets
36	Sideplate sub-assembly - R.H.
37	Chassis strengthener sub-assembly - B14
	Chassis strengthener sub assembly - B17
	Chassis strengthener sub-assembly - B20
38	Engine platform sub-assembly - B14
	Engine platform sub-assembly - B17
	Engine platform sub-assembly - B20
39	Bolt for engine fixing - front
40	Screw for engine fixing-rear & platform fix.
41	Screw for platform fixing-rear & strengthener
42	Screw for chain cover pillar & delivery plate
43	Washer for chain cover pillar
44	Grassbox sub-assembly - B14
	Grassbox sub-assembly - B17
	Grassbox sub-assembly - B20

CHAPTER EIGHT

MAINS ELECTRIC AND CORDLESS MOWERS

When operating any type of outdoor electrical product it is always advisable to use an RCD (residual circuit device); this useful inexpensive safety device just plugs into the mains socket and then the machine is plugged into that. It can be used on any electrical machine inside or outside and it is designed to protect against electric shock, as it can be so easy to cut through the cable of a lawnmower, hedge cutter or chainsaw.

Before doing adjustments or service work on any electrical product, always remove the plug from the mains supply.

Managing the cable

Mower mains cables can easily end up horribly twisted, becoming a nuisance to unravel every time you get the mower out. If you notice the cable starting to curl it means you are twisting the cable either when you wrap the cable up, or that the sequence in which you mow the lawn puts one twist in the cable each time you turn or store the cable. One of the secrets here is to ensure the cable is unplugged when wrapping the cable for storage. Besides being safer, it

LEFT: The first mains electric lawnmower patented in 1926 by Ransome, weighing over 2cwt. The switch mechanism was made of wood and copper. A counter-balance weight was fitted, which ensured a perfect roll and stripe, as the motor was not fitted perfectly centrally. Note the swing bar, which protected the main cable.

BELOW LEFT: Residual current device (RCD); protects the operator against electric shock.

BELOW RIGHT: Mowing guide: direction of cut for a traditional striped finish with a mains cable.

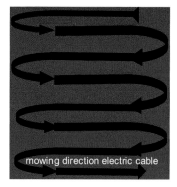

mowing direction electric cable

Wiring for a 13amp mains plug, with the Earth connection to the large centre pin. Most modern double-insulated electric mowers do not require the Earth wire.

allows the cable to untwist when wrapping each time around the cable holders or around the handle.

When mowing with a mains cable, start at the position nearest to the mains supply with all the cable unwound, and then always travel in a direction away from that point; this sequence will ensure that the cable will always be on the side you've already cut, making it difficult to cross back over the cable. On the turns, always turn away from the start point so you avoid the cable crossing over the machine, this also prevents the cable from twisting. If the cable is badly twisted, worn, cut, or damaged in any way, it's best to replace the whole cable. Under current regulations it is unsafe to repair a cable with a join.

When using extension cables, ensure the cable is of the correct specification for the machine; depending on the model, we would recommend a minimum of a 10amp cable on low power models; ideally 13amp 3,000 watts spec. is best with 1.25mm cable. This thickness is strong and will cope with the drag over the grass. If too low a spec. is used this can cause a reduction in power to the motor, which may cause the extension cable, the plugs, or the motor to overheat. Always unwind the cable, as if it is left wound on the reel the cable can act as an electrical coil and

it will heat up more quickly. Check on your reel, as it will have a maximum wattage, ampere rate and cable thickness stated on it. If your cable total is over 100 metres long, a power drop can occur, and a cordless or petrol model may well be worth considering. A cable of this length can also become quite inconvenient. Lastly, do not use a mains cable or extension cable in the wet.

Mains plugs on modern electric mowers are now mainly moulded and sealed to the mains cable. This is to stop moisture and prevent the plug being tampered with. But there are times when the plug may get damaged and a replacement is required. If your machine is double insulated, an earth wire is not required. The brown wire is connected to the positive terminal with the fuse, the blue wire to the negative terminal. If the cable has a third yellow or green wire this is connected to the central earth pin. Do not fit a higher-rated fuse than the one originally supplied; it is there to protect the machine. If you are not sure, consult your handbook or your authorized service agent for the correct size. Fuse sizes can vary depending on the motor type and wattage; each electric mower has a data tag with the wattage power rate.

Note: when carrying out any electrical work, only attempt the job if you are a hundred per cent confident, otherwise give the job to a qualified person.

When you take your electric machine into any authorized service agent for service or repair, after the service each machine will be checked on a 'Safety Flash Tester'. The job will be logged and issued with a registered number, which is recorded. The test is checked with a 'pass' or a 'fail'. Should a fail be indicated the fault would be found and rectified. The electrical flash test checks if there is any electric leakage anywhere on the machine (so it is electrically safe for you to touch whilst it is plugged into the mains supply without danger of an electric shock) and it is the final check before the machine is returned to the customer. It gives peace of mind to you and the servicing agent and ensures that the machine is electrically safe to use.

Note: when folding the handles on a mains electric mower for storage or transportation, check that the switch cable to the motor is clear of the handle pivot bolt. If the cable is trapped inside the handle at this point it is highly likely to damage the cable insulation or sever the wires, creating an unnecessary repair.

MOTOR WATTAGE SIZES AND FUSE RATINGS ON SOME COMMON ELECTRIC MOWERS

Alko 34E Induction motor	1200 watts	13 amp fuse
Alko 40E bio Induction motor	1400 watts	13 amp fuse
Atco Consort Brush motor	340 watts	13 amp fuse
Atco Regent Brush motor	1200 watts	13 amp fuse
Atco Richmond 36 Induction motor	1300 watts	13 amp fuse
Atco Windsor Brush motor	340 watts	13 amp fuse
Bosch ASM 32 Brush motor	340 watts	13 amp fuse
Bosch Rotak 32 Brush motor	1000 watts	13 amp fuse
Bosch Rotak 34 Brush motor	1200 watts	13 amp fuse
Bosch Rotak 36/37 Brush motor	1200 watts	13 amp fuse
Bosch Rotak 43 Brush motor	1400 watts	13 amp fuse
Flymo Glidemaster Brush motor	1650 watts	13 amp fuse
Flymo Hovervac Brush motor	900 watts	13 amp fuse
Flymo Microlite Brush motor	1000 watts	13 amp fuse
Flymo Roller Compact 4000 Induction motor	1600 watts	13 amp fuse
Flymo Turbo Compact 300 Brush motor	1500 watts	13 amp fuse
Flymo Turbo Compact 380 Brush motor	1700 watts	13 amp fuse
Flymo Turbolite Brush motor	1000 watts	13 amp fuse
Flymo Turbolite 350 Brush motor	1500 watts	13 amp fuse
Flymo Turbolite 400 Brush motor	1500 watts	13 amp fuse
Hayter Envoy Induction motor	1400 watts	13 amp fuse
Hayter Harrier Induction motor	1500 watts	13 amp fuse
Mountfield 320 Induction motor	1100 watts	13 amp fuse
Mountfield 390 Induction motor	1500 watts	13 amp fuse
Mountfield 410 Induction motor	1500 watts	13 amp fuse
Qualcast Classic Brush motor	340 watts	13 amp fuse
Qualcast Concorde Brush motor	280 watts	3 amp fuse
Qualcast Concorde/Elan Brush motor	400 watts	13 amp fuse
Qualcast Electric Punch Brush motor (early models)	340 watts	3 amp fuse
Qualcast Turbo 320 Brush motor	1000 watts	13 amp fuse
Qualcast Turbo 340 Brush motor	1200 watts	13 amp fuse
Qualcast Turbo 400 Brush motor	1400 watts	13 amp fuse

Cordless machines

Of course cordless electric machines don't have any of the above issues, as the maximum voltage at present is either only 12, 24, or 36 volts (compared to 240 volts on mains power), so there is no danger of any electric shock and there are no trailing cables. (Remember, however, that the blades are just as sharp.) Modern cordless domestic machines are light and are the easiest and most convenient to use: no petrol, no oil, no rope pulling, no starting issues, and no cables. The main difference here is the battery, which requires recharging, easily done on modern cordless mowers, especially if fitted with the new technology Lithium ion type. These batteries are lighter, have more energy, have a longer life and have no memory so can be charged at any state. (Older batteries, if not discharged fully prior to charging, would remember the unused portion of charge and after recharging would not allow this portion of the battery to be used again. This becomes known as the battery memory and results in the useable portion of the battery becoming smaller. Newer batteries with no 'memory' do not suffer from this and when recharged do not require a complete discharge first.) They can achieve more than 75 per cent of their charge in a very short time; coupled with a spare battery, this will cut fairly large areas.

Vintage battery electric cylinder lawnmowers of the 1970s used a 12 volt car-type battery, and have stood the test of time – some are still being used today. These machines also had the advantage of being self-propelled and have been popular over the last forty years.

Modern cordless mowers now come with cylinder and rotary type blades. Some cylinder types can still cut the grass even when the battery is completely flat. On some models the same battery can be attached to other appliances, for example, hedge cutters, grass trimmers, powered pruners etc., making them quite versatile.

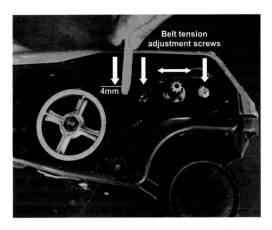

Adjusting the toothed drive belt on a Qualcast Concorde, Elan and Eclipse. Slacken off the two large cross-head screws half a turn anticlockwise; move the small drive pinion forwards or backwards to tension the belt. Ensure the adjustment screws are retightened. Test by moving the belt up and down – the amount of movement should be about 4mm.

Servicing and repair

On mains-powered and cordless machines most motor servicing is best left to a professional.

The cutter mechanism is virtually the same as petrol and mains electric models. To maintain the sharpness and balance on mains or cordless rotary blades, follow the same procedure as in Chapter 5 (*see* page 46). When removing an electric rotary blade, check the blade bolt for any cracks or damage, and replace the bolt if worn. The outer casing on these bolts is often made of some type of plastic; besides holding the blade in place, it gives insulation to the blade. It is therefore important to use the correct size spanner so not to cause any damage. If the blade is worn, damaged or bent it is best replaced; if in good order it can be sharpened and balanced, or take it to your local service dealer for a regrind and

Qualcast electric Punch drive mechanism. It includes a tooth belt, V-belt and two chains.

Atco Windsor and Qualcast Classic QX mains electric V-belt drive mechanism. It includes two V-belts and a tooth belt.

balance. Regrinding or new electric rotary blades are not prohibitively priced. Once balanced, it will take any excessive vibration out of the machine, which makes it less fatiguing to use, and will give the machine a much longer life.

There are motor filters on electric mowers, and depending on grass conditions, the filter and motor will eventually clog up with grass and debris. When clogged, the motor will heat up and prematurely wear. In the case of hover mowers, the machine can lose its hover-ability and power. An annual motor service is recommended to keep your motor going, especially if it is being used in dry, dusty conditions.

If you need any servicing work or spare parts for an electric Flymo, you will need the serial number and the product number. These can be found on every Flymo on a silver label fitted on the deck at the rear or side. Part of the serial number can decipher the date, year and month of manufacture, and it can come in handy if you have lost your receipt, especially if work is required under the manufacturer's warranty. The Flymo company has a huge network of service dealers all over the world, especially in the UK.

If you need any service work or spares on Atco, Suffolk and Qualcast machines, quote the serial numbers and product numbers, which also identify model and product type.

TOOTH BELTS

These belts are extremely strong and hardwearing; they can last virtually the life of the machine. If you find a tooth belt broken or snapped, the most likely cause is that the blade has had an impact. Check the blades before refitting a new belt.

CHAPTER NINE

GARDEN RIDE-ON TRACTORS

Garden tractors are a huge subject and could more than fill this book just on their own, as there is such a tremendous variety of models and manufacturers on the market. The main advantage of owning a garden tractor is that as the machines have a wider cut and can travel faster than walking speed, they will take far less time. The machine will last longer as a consequence, as it will be doing fewer hours and miles than its walk-behind counterpart, notwithstanding the fact that you are sitting down whilst mowing, having only turned a key to start – and you don't even have to leave the seat to empty the grassbox.

Domestic ride-on models have engine powers from 5hp to 26hp. Widths of cut range from 24 inches to 54 inches, with either a single blade, twin blades or triple blades, depending on the size. The information in this chapter is valid for the five main types of ride-on domestic garden tractor – side eject, direct collect, rear discharge powered sweeper, mulch and front deck rear steer.

Side eject models are suited for longer and rougher grass areas. The grass is quickly and efficiently discharged straight out of the side of the cutter deck. These models are ideal if the grass is not required to be collected. Some side eject tractors can easily be converted to mulching mode with a mulching kit, or to grass collection via a tube to removable grass boxes at the rear; this has the advantage of the grass being able to be emptied into a green wheelie bin or to be tipped on top of a compost heap.

Direct grass collect tractors are equipped with a simple rear pivoting grass box that can easily be emptied without the operator leaving the seat. Many of these models can be quickly converted without tools to mulching or to deflecting the grass to the ground if the grass is not required to be collected. They are efficient and convenient to use.

Rear discharge with a powered sweeper models work by cutting the grass and leaving it on the ground, closely followed by a revolving brush sweeper driven by a belt via a power take off (PTO) from the gear box. The grass can be emptied without the operator leaving the seat, and many of these collectors are fitted with a roller to give a groomed striped finish.

Mulching tractors are dedicated to just mulching and in general are very good at it, the big advantage being never having to collect the grass. Moreover the finely chopped grass acts as a fertilizer to feed the grass roots, leaving it quite lush; the nitrogen-rich clippings naturally fertilize without chemicals.

Front deck models, more commonly known as 'Riders', have the advantage of a much tighter turning circle where manoeuvrability is important. This is because rear wheel steering makes them extremely manoeuvrable in tight spaces. The cutters are mounted in front of the driving wheels, enabling awkward areas such as corners and under low bushes to be cut. Most front deck riders are dedicated mulchers – an advantage over conventional tractors that collect and mulch which sometimes can be a compromise.

Poor grass collection

Several things can cause poor grass collection problems. Check that the grass chute and the deck are not blocked or clogged. Check that any air vents on top of the deck are clear. Remember that wet or damp grass sticks to everything, and always try to cut when the grass is dry. Check the condition of the blades – are they damaged? Check the blade wing tips for wear; these create the draft to throw the grass into the box. Check the blade sharpness and balance – is the deck vibrating more than when you originally bought it?

If these quick checks are ok and the problem still occurs, raise the height of cut slightly and or cut at a slower speed. Check that the engine is functioning smoothly; this could simply be a clogged air filter which will cause the engine to run more slowly and lose power. Check the air filter every twenty-five hours, more often if running in dusty conditions.

Remember the Golden Tractor Rule: the slower the speed the better the cut! Never let the cutters or engine labour, this will cause the drive belt to heat up and prematurely wear.

Brake problems

A seized brake can have the symptom of no drive; so don't be fooled as the problem can often be easily sorted quickly. Most tractor brakes are situated on the side of the manual or hydrostatic gearbox, consisting of a 2–3in steel disc that spins when put into drive. The disc is sometimes hard to see as it can be partially shrouded by the wheel, brake pad, brake linkage or chassis. A seized brake can be caused by leaving the brake in park position when the tractor is not in use, when moisture can cause the disc and brake pad to oxidize and seize together. When storing the tractor, leave it without the brake engaged. If the

brake should seize, put the gear lever into neutral if you have a manual gearbox model; if it is hydrostatic, disengage the hydro drive (normally via a lever near the gearbox). Try rocking the tractor back and forth till the brake disc moves. If this doesn't release it, look for the brake linkage, following it from the brake pedal to the brake pad. The linkage pushes the pad onto the disc, so try pushing the linkage back toward the pedal, then try rocking the tractor again; nine times out of ten it will work. If not, remove the rear wheel nearest to the brake and tap the brake. If this fails, call your service dealer.

Punctures

Starting with punctures, getting one wherever you are is always a pain in the 'gr-ass' (pardon the pun) and the saying 'it's only flat at the bottom!' doesn't help, as a puncture will mean the tractor won't steer and the tyre will be damaged if driven on. But if you have anything like a thorn hedge in your garden there's nothing more certain that sooner or later you're going to get a flat. Thorns are extremely sharp and strong; they will go

THE LAW OF THE GARDEN TRACTOR

- It will always puncture in an awkward place, and where it will be extremely hard to push.
- It will be the farthest place away from the shed.
- It will rain as soon as you get a flat.
- It will be in the evening or the weekend when the shops are shut.
- The grass will not have been cut for three weeks, and you will be going on holiday the next day.

LEFT: **Fitting tyre sealant eliminates punctures and flat tyres, saving time and money. If you have any thorn hedging its advisable to have tyre sealant fitted on the annual service, or ideally ask for it to be fitted before you get delivery of the tractor.**

RIGHT: **Tyre valve removing tool, often supplied with bottle of tyre sealant.**

straight through a garden tractor tyre and the inner tube if one is fitted (most domestic tractor tyres are tubeless). But it doesn't have to be as annoying as you think when you get one.

There are several ways people tackle this; the most common, which amazingly, is the most time consuming, most expensive and least effective is to jack the tractor up, remove the wheel, take it to a car tyre dealer to have the puncture repaired – or wait for them to order an inner tube as they do not stock tyres that small – return and refit the wheel only to pick up another thorn on the way back to the shed.

There are several tyre sealants that will cure the problem of punctures permanently. To fit tyre sealant is an easy operation:

1. Position the wheel with the valve at the four or eight o'clock position.
2. Allow the tyre to deflate: remove the centre of the valve by unscrewing the internal valve stem anticlockwise with the valve core remover (normally this is provided with the bottle of sealant).
3. Fit the plastic tube provided with the sealant onto the outside of the valve and squeeze the sealant slowly into the tyre.
4. Check how much sealant is required by reading the tyre size embossed on the outside of the tyre and the quantity of sealant stated on the bottle. (A rough guide, depending on the size of the tyre, is that approximately 1 litre of sealant will cover two domestic garden tractor front tyres. Larger tyres will require more sealant to give adequate coverage.)
5. Refit the valve stem by carefully screwing the valve stem clockwise.
6. Spin the wheel and tyre several times to ensure the sealant has covered all the inside of the tyre.
7. Inflate the tyre to the correct pressure (many tyres have the pressure reference stamped on in psi or bar), and you will have no more punctures in that tyre. If you pick up another thorn or puncture the sealant automatically will reseal the hole.

Tyre sealant can also be used on wheelbarrows, sack trucks, golf karts, garden tractor-trailers and mobility vehicles, etc. There are also temporary emergency tyre inflator canisters available. These convenient compressed air canisters enable you to re-inflate the tyre without removing the wheel.

Blades

The blades on garden tractors need to be checked regularly and reground and balanced every season – or, in the case of a blade having been impacted, straight away. To inspect the cutters and shafts properly, it's best to remove the cutter deck assembly unit. This may not be as hard as it looks and generally will only take a few minutes. There are many different designs of cutter deck but most domestic tractor models have the same principle to remove the cutter unit. This instruction is a general guide only; your model may differ, so always consult your handbook.

To remove the cutter deck of a side discharge tractor:

1. Remove the side chute.
2. Put the cutting deck to the lowest cutting position.
3. Ensure the cutter engage lever is in the disengaged position and the cutter belt is slack.
4. Gently ease the cutter belt off the engine pulley.
5. Remove the R clip(s), pin(s) or bolt(s) from the front of the cutter deck (they may be on bars that go to the front of the machine).
6. Remove the two R clips and pins or bolts from the rear of the cutter deck.
7. The cutter deck should now be resting on the floor. If it is not, check to see what is still stopping the cutter deck from resting on the floor and disconnect it.
8. Disconnect any electric wires that are connected – these are usually fixed by a push-on electrical connector.
9. Pull the cutter deck out by sliding it from underneath the tractor. It is not necessary to remove the cutter belt off the deck unless it needs replacing. If you find the deck snags on something when sliding it out you may find it easier removing it from the opposite side.

Once the deck is removed, clean off any grass or debris from the top of the deck with a stiff hand brush or garden hose. Inspect the cutter belt for any wear or cracks. Check the belt pulleys for any distortion or wear, and make sure that the pulleys spin true. Check that the pulleys are not loose by checking the central pulley nut. Check for any play in the cutter shaft bearings. Check the wear on the pulley brake disc, if fitted. Normally the cutter belt would be changed each season; before removing the belt, make a note of its configuration around the pulleys.

Having checked the top of the deck, carefully turn it upside down. Check underneath the metal deck for any corrosion, especially around the blade mountings. If there is heavy corrosion in this area, let a professional look at it. Most tractor cutter blades bolt onto a central blade-mounting boss with a single or two fixing bolts. The bolt(s) are attached to a drive shaft, which is mounted in aluminium or steel housing often called a 'quill housing'; this is bolted onto the tractor cutting deck. The important thing here is to check the quill housing and shaft to see if the bearings are in good order. Wear gloves, as these blades can be sharp.

1. Spin the blade and check if it runs smoothly without any grinding or 'notchy' noises.
2. Check the blades for any damage or wear.
3. Check that the blades are level and not bent by turning the blade through 180° to see if the opposite tip is at the same height.
4. On twin or triple blade decks check that all the blades are at the same height when they meet each other in the centre.
5. Grip one end of the blade firmly and rock it up and down. If there is any movement or play in the shaft the bearings require replacement (on some models the bearings, drive shaft and housing come as a complete unit).

When removing the blades, most blade bolts are right-hand thread. But on some twin contra-rotating blades one blade will have a left-hand thread. It is very important to regrind and balance each blade, as an unbalanced, bent or dam-

Plan view of a popular cutter deck, showing belt configuration. Mountfield 1436 tractor.

Underneath view of the cutter deck, showing blade alignment. Blades must be level when they meet otherwise an uneven cut will occur.

aged blade will give an uneven cut, create uncomfortable cutting, have a slightly different noise, and will create premature wear on the whole of the tractor. It is advisable to take tractor blades to a specialist for regrinding and balancing, as some tractor blades have a shaped centre hole, making it difficult to find the centre. Blades can also be longer, with special shaped aerodynamic wings to assist grass collection or create mulch. When refitting tractor blades after they have been checked, make sure that the blade wing tips are facing upwards. Ensure that the

blade bolts are tight. Reverse the procedure on page 71 to refit the cutter deck. Make sure that the belt is fitted the correct way on the jockey pulleys before starting the engine. Especially check that the belt is on the correct side of any belt guides.

Can't start, won't start

Before starting, ensure that:

- The fuel is switched on.
- The brake is depressed.
- The parking brake is on.
- The cutter is disengaged.
- The grassbox is fitted correctly.
- You are sitting on the seat.

Electrical problems

One of the main common and niggly causes of a tractor not starting can be a simple electrical fault. There are several safety switches; if any of these are out of adjustment, or have been contaminated with dirt or debris, or are faulty, the engine will not be allowed to start. These switches are fitted to avoid a potential accident: they come into play every time you sit on the seat, press the brake pedal, engage the cutter, shift to reverse gear, or when emptying the grassbox.

The switches are situated:

- Under or in the seat.
- On the brake pedal or linkage.
- On the end of the cutter engage lever or cutter switch.
- On the rear of the machine where the grassbox contacts the frame.
- On the reverse position on the gear lever.

Check each switch individually, making sure that it is depressed when the appropriate operation is engaged. Check the grassbox switch first as this is the most likely to alter its adjustment, especially if you've caught the box on an object.

Depending on the model, some tractor blades will not work if the grassbox is not positioned properly. Some models will not cut whilst reversing; others have a reverse cutting switch.

Fuel problems

Fuel also plays a big part in start-up on some engines. To stop debris going into the engine an in-line fuel filter can be inexpensively and easily fitted in seconds. Many in-line filters have a transparent outer casing so the amount of fuel and debris can be seen.

For other possible explanations for your tractor failing to start, *see* Chapter 11, Fault Diagnosis.

The tractor seat safety switch stops the engine if the operator leaves the seat without engaging the parking brake.

In-line fuel filter for walk-behind or ride-on garden machinery.

Tractor foot brake safety switch.

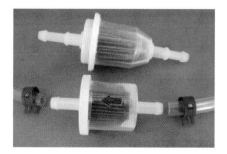

In-line fuel filters for garden tractors, showing direction of fuel flow.

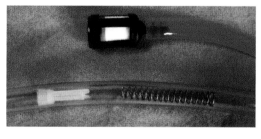

In-line fuel filter fitted on lawnmowers, and tank pick-up filter, found in hand-held power equipment such as chainsaws, hedge cutters, brush cutters and blo-vacs.

In-line fuel tap and fuel tap with filter.

RIGHT: Tractor ignition switch. Replacing an ignition switch is a relative easy job, but there are at least ten different ignition switches, all of which appear similar. Using the wrong switch can cause the wiring to short. Ensure that it has the matching terminal connection configuration on the back of the switch.

CHAPTER TEN

ROBOT MOWERS

In the distant past Ransome and Webb lawn-mower companies pioneered and produced remote controlled robot-like lawnmowers, developed in the 1950s and 1960s. Webb based their lawnmower on a 14in cylinder lawnmower powered by a 2hp Briggs & Stratton petrol engine which powered the rear roller, but fail-safe technology had not quite been perfected then, and the machine could be prone to cutting a swathe through the flowerbeds and then would proceed through the next-door neighbour's garden fence.

The first mass production robot mower was introduced in 1995, manufactured by Husqvarna, a Swedish company which is part of the huge Electrolux conglomerate. It was powered by solar panels, only requiring daylight for the robot to work. The robot mower is virtually silent in operation unless it is stolen, when it emits an alarm; it also has a security pin code, which has to be entered before it can be used. This robot works like a sheep and wanders around the lawn, and when it senses where the grass is slightly longer it works in that area before going on to the next. The first robot cost £1,000,000 pounds to develop and models were sold in the shops for £2,000 each. There are now several different kinds of robots on the market, and they have become less expensive. Some cut in straight lines and are powered by battery; when a robot senses that the battery is running low it will go back to its docking station to recharge itself – when recharged it will then go back out on its own to

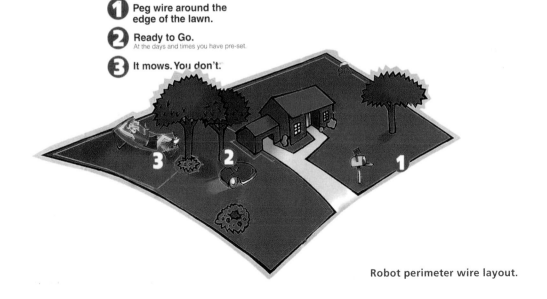

1 Peg wire around the edge of the lawn.

2 Ready to Go.
At the days and times you have pre-set.

3 It mows. You don't.

Robot perimeter wire layout.

cut again. One advantage of a robot running from a battery instead of solar panels is that the robot can be programmed to cut the grass in the pitch dark at midnight or on any day or time of your choice, come rain or shine. In the near future, robots cutting grass may be a very common sight. There are already big powerful petrol robots looking like monster spiders, which are cutting undulating surfaces over thousands of acres of grass such as on golf courses, embankments, water reservoirs and verges, on all sorts of terrains all over the world.

Setting up

To set a robot up, a low-voltage perimeter wire is laid out all around the outside of the lawn area. If there are trees, ponds, flowerbeds etc. to be avoided, the wire can be extended around the obstacle. Safety is a big feature on robot mowers. If the robot should encounter an animal or child, the robot's safety mechanism kicks in and is automatically programmed for the blades to stop (within less than a second), the robot will stop until the hazard has gone, or it will proceed in the opposite direction, depending on the model purchased. Other in-built safety mechanisms are magnetic and optical; in these, the blades automatically stop rotating if the robot is lifted up or tilted. Some robots have a remote control feature so they can be guided from one area to another or in awkward areas. Battery robots use sealed batteries, which will not leak fluids regardless of position. Current robots on the market will not operate outside the perimeter wire: it creates a virtual wall visible only to the robot, and keeps your robot where you want it, on the lawn. If the safety perimeter wire is not working, neither will the robot! The perimeter wire can be laid on top of the ground or can be buried up to 10cm (4in) deep. Small pegs are supplied with the robot, which are designed to fasten and hold the perimeter wire to the ground or below grass level. (It is a lot easier to insert pegs and wire if the ground is moist.) Once fastened down, the pegs will soon disappear under the growth of new grass and will not be visible. Once the battery is connected, the robot will wake up, and give instructions via its LCD display on how to complete the set-up. If a fault occurs, the LCD display panel will instruct the owner to perform a certain function or action.

All current robots are fitted with a rotary cutting blade. Cutting blades on all robots are very easy to change, and can be re-sharpened or quickly replaced. One thing the robot cannot do is check the lawn for any stones or debris before it starts, so this is one job for you!

Winter storage

Remove the battery fuse and clean the robot. Reinsert the fuse and connect to the mains supply. The charging system is designed such that it can remain plugged in at all times without any danger of over-charging, over-heating or damaging the battery. Store indoors in a clean dry place on its wheels. Disconnect the power supply to the perimeter wire.

If after reading this book you have still not been converted to join the contented half of the country who are mower-happy people, and still think your mower hates you, your saviour may be a robot mower. These machines are specially made for people who want to be in charge of cutting the grass but do not want to mow it. (Be warned: don't let the robot mower into the house, as it will make short shrift of the shag pile carpet.)

Finally, if you are still not convinced – you still want nice short grass rather than an overgrown garden, but you still don't want to mow – you're best option is to give up and get a goat!!

The world's first robot mower, with a production cost of £1 million. Manufactured by Husqvarna in 1995, it is powered by daylight and wanders round the garden like a sheep.

The Robomow, manufactured in Israel. This type of battery-powered robot cuts in straight lines, and returns to its docking station to recharge when it senses its power is low.

Ransome Jacobson Robot Spider. This powerful machine is uniquely remote controlled with a hydrostatic drive and dedicated to cutting grass on slopes. It will operate on inclines up to 40° (or 55° with its onboard winch) and can mow where no man has mown before, incorporating four-wheel drive and four-wheel 360° steering, and an 80cm or 1.23m width cut that can be mown in any direction. The robot is more productive than six men with petrol trimmers.

CHAPTER ELEVEN

FAULT DIAGNOSIS

The spark plug

A quick and easy method of diagnosing why the engine will not start is by removing the spark plug. The condition of the spark plug tells a lot about how the engine is running and can be a good indicator as to why the engine is not starting.

Before removing the spark plug, however, check the spark plug cap. If it has a sloppy or loose fit, the connector terminals inside may just need squeezing closer together for a snugger fit. To make a snugger fit on some types of plug cap a screwdriver can be placed inside the cap; gently push the terminals together for a closer fit for a better connection.

If there is no spark at all, whatever you do with the spark plug or cap the engine is still not going to start. The spark plug itself can be given a simple spark test. There is a very handy inexpensive little tool made by Briggs & Stratton for this purpose, which can be used on virtually any petrol engine. Simply remove the spark plug cap, clip one end of the tester on the spark plug, the other on the end of the cap to test if there is a spark. Crank the engine over and check if there is a spark. (Note: if the daylight is too bright place a piece of black insulation tape over one side of the spark tester – the black background will make the spark easier to see.)

Different engines require different types of spark plugs: engines that tend to run hot need cold plugs; those that run cold demand a hotter type. The plug's heat range determines the spe-cific plug for any engine, that is, the maximum and minimum temperatures between which the spark plug will give optimum performance. Good quality proprietary spark plug brands, with copper wire instead of iron cores, have more resistance to fouling and have a higher pre-ignition rating. The copper dissipates heat quicker because of its superior heat conductivity and cools the electrode and insulator tip. Wide heat range spark plugs are more flexible; they perform equally well in hot or cold engines (at high revs or low revs, stop and go, or continuous running such as in generators, etc.). Increased heat resistance does not affect fouling resistance, which is primarily determined by the insulator nose length. The longer the nose, the more susceptible it is to heat and the freer from fouling.

Always check the condition and cleanliness of the spark plug, especially the spark plug sealing

Spark plug tester, Briggs & Stratton part no. 19368.

'Some people swear by other plugs' – a cartoon from NGK, spark plug specialists.

ened it will distort. When tightening head bolts tighten opposite bolts in sequence (there are many different torque settings – it is worth consulting a workshop manual for the correct torque setting for your engine). On some engines access is awkward; if this is the case, on refitting a spark plug place a piece of rubber tubing over the plug and install by hand, then lock up using the correct spark plug spanner which will not damage the plug or engine cylinder head cooling fins. Don't forget to check the spark plug cap and high tension lead condition.

If 'R' is in the spark plug number this indicates a radio interference resistor is fitted. If 'LM' is in the spark plug number this generally indicates it is designed for a lawnmower.

When replacing a spark plug, before fitting, ensure the replacement spark plug base, electrode and threads are no longer than the original as damage to the engine internals may occur if a

ring and the threads in the cylinder head. If the cylinder head threads are damaged (usually caused by cross threading, or over tightening the plug), either replace the cylinder head, or the threads can be repaired by fitting a helicoil. A helicoil is a repair kit that replaces damaged threads back to the original size. It is particularly useful if the spare part is not available, is too expensive to replace or if there is not enough room to fit a larger plug or bolt. It is worth checking the price first to find out which would be the most economical, or ask your service dealer: they may have a good and inexpensive used cylinder head.

When refitting a cylinder head, always replace the cylinder head gasket. Do not refit a cylinder head when it is hot: as most modern lawnmower cylinder heads are made of alloy, the hotter it is the softer the metal, so when the bolts are tight-

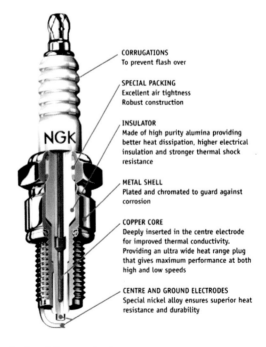

CORRUGATIONS
To prevent flash over

SPECIAL PACKING
Excellent air tightness
Robust construction

INSULATOR
Made of high purity alumina providing better heat dissipation, higher electrical insulation and stronger thermal shock resistance

METAL SHELL
Plated and chromated to guard against corrosion

COPPER CORE
Deeply inserted in the centre electrode for improved thermal conductivity. Providing an ultra wide heat range plug that gives maximum performance at both high and low speeds

CENTRE AND GROUND ELECTRODES
Special nickel alloy ensures superior heat resistance and durability

Parts of the spark plug.

longer plug is fitted. Most lawnmower spark plugs have a short thread (except some Honda engines), unlike car spark plugs which usually have a long thread.

It is important to have the correct electrode gap. An incorrect gap can cause bad starting and poor engine running. Check your engine handbook for the correct setting. It is worth checking

SPARK PLUGS FOR LAWNMOWERS

Listed below are some popular lawnmower spark plugs and comparison equivalent numbers. (This is just a small selection and is not comprehensive, but does cover some very common mower engines.)

Spark plug make and model	Engine make	Spark gap
Champion J8, J8C, J17LM NGK B6S Bosch W7E, W9ECO	Briggs & Stratton, Qualcast, Suffolk, Tecumseh, Aspera	0.6mm
Champion J19LM NGK B2-LM Bosch W11E0 Denso W9LMUS	Briggs & Stratton, Tecumseh, Barrus, various Chinese engines	0.8mm
Champion RJ19LM NGK BR2-LM Bosch WR11EO Denso W9LMRUS	Briggs & Stratton, Tecumseh, various Chinese engines	0.8mm
Champion RN11YC, N11YCC NGK BPR5ES Bosch WR8DC Denso W16EPRU	Honda	0.7mm
Champion L86C, L86CC NGK B6HS Bosch W8AC Denso W14FU	Various Villiers and Chinese engines	0.5mm
Champion L10, L86C, L86CC	Various Villiers, four-stroke engines	0.6mm
Champion D16, 8COM	Various Villiers, two-stroke engines, also older vintage models (18mm diameter thread)	0.8mm

Testing the electrode gap on a spark plug with a feeler gauge.

the spark plug gap regularly, even on a brand new spark plug – if a new plug has been roughly handled or transported the gap may have altered. Remove the spark plug and check the electrode gap.

There are hundreds of different sizes, shapes, heat ratings, spark plugs all with different part numbers. These are determined by:

The electrode is in good condition, so the engine will run correctly. The insulator should have light grey or brown deposits. Even if the spark plug is used under good conditions carbon deposits will still accumulate on the electrode. Regular inspection is advisable; clean or replace the spark plug when necessary.

- Thread diameter size (8mm, 10mm, 12mm, 14mm, 18mm).
- Thread pitch size (1mm to 1.5mm).
- Hex size (13mm to 25.4mm).
- Taper seat (7.8mm to 20.8mm).
- Thread reach (9.5mm to 19mm).
- Spark gap pre-set (0.8mm to 1.5mm).

The condition of a spark plug electrode can indicate how the engine is running. On an overheating spark plug, the insulator is white or blistered. If the temperature rises above 870°C (ideal heat range is between 460°C and 860°C), pre-ignition may occur. The engine will lose power and major internal parts of the engine can be damaged especially the piston.

The causes of overheating can be:

- Incorrect fuel mixture with insufficient octane (old or stale petrol).
- Fuel mixture is too lean or weak (too much air and not enough fuel).
- Blocked cooling fins or fan.
- Engine cowling clogged with debris.
- Over-advanced ignition timing.
- Excessive carbon deposits in the cylinder head (this can be easily rectified by an engine de-coke).

If the spark plug core is split, this can possibly be a hairline crack caused by over-advanced ignition, insufficient petrol octane or weak mixture; check the cooling system. The spark plug requires replacement.

An air-cooled engine will run very hot if run on old, stale or contaminated fuel.

If the plug is dry, it is most likely to be a fuel problem. Fuel is not getting into the engine cylinder head; there could be several causes for this, from something extremely simple to something quite technically involved. Start by checking the easy things first; they will only take a few minutes.

RIGHT: Oil-fouled spark plug. Oil has accumulated on the spark plug nose, which forms a short circuit or earth leakage path. The spark is eliminated or weakened; the engine will misfire, causing bad starting and poor acceleration. The causes are: worn valve guides, piston rings or engine cylinder bore, or it could be overfilled with oil. Remedy: check oil level, temporarily replace spark plug to get the lawn finished, then have the engine checked by a professional.

LEFT: Carbon-fouled spark plug. Carbon deposits have accumulated on the spark plug nose, forming a short circuit or earth leakage path; this can be common with unleaded petrol. The causes are: over-rich mixture (too much petrol), faulty choke mechanism, or clogged air filter. Remedy: check the air filter, choke and carburettor setting. Heavy deposits on the spark plug could be caused by too much upper cylinder lubrication, possibly by worn valve guides. Replace the spark plug.

RIGHT: Damaged spark plug. An incorrect plug has been fitted, damaging the plug threads and electrode. The engine piston and valve may also be damaged. Ensure that the thread length is no longer than the original when replacing a spark plug, otherwise permanent engine damage may occur.

No-start checklist

- Check there is fuel in the tank (this sounds obvious, but it is often overlooked).
- Check that the fuel tap is switched on.
- Check petrol is coming out of the fuel pipe, preferably by removing the end fitted to the carburettor; if this is not easily accessible, remove the pipe at the tank end to see if fuel is coming out of the tank.
- Check the throttle control cable to see if it is moving and travelling to its full extent via the control lever on the handle bar.
- Check if the choke lever is moving to its full extent (this moves a circular brass disc called a butterfly inside the carburettor which swivels on a brass spindle); quite often, depending on the engine, this can be seen moving when the air filter is removed.

Neglected spark plug: moisture ingress and corrosion. Replace plug.

- Check the primer bulb (if fitted).
- Is the petrol fresh, or is it more than six weeks old?
- Is there fuel stabilizer in the fuel?
- Check if the engine on-off switch is on; sometimes the switch wires can short.
- Check the operator presence control (OPC) lever and control cable. If the OPC is out of adjustment there will be no ignition. (On some engines if it is quiet, you can hear the switch click on by pulling up the OPC slowly when the engine is stopped.)
- Check the spark plug cap.
- Check if the choke is working. If the engine will not start because it has flooded (too much fuel in the cylinder head) it is best to go for a cup of tea for ten minutes, then try starting it.

If after these simple checks have been carried out, there is still no joy, the carburettor may need stripping and checking.

If the spark plug is wet it is most likely to be an electrical problem, as there is fuel getting to the plug, but it is not igniting the fuel for some reason.

It is always worth buying a spare spark plug when purchasing a new mower (ask if the shop will throw a spare spark plug in with the new mower – there's no harm in asking). Or ask for a spare plug on its annual service. It's easy to change a spark plug and if you've got a spare on hand, it can save a trip to the shops, saving valuable barbecue time on your sunny afternoon.

Note: if the engine keeps stopping after every five to ten minutes, but then restarts each time after a few minutes, check that the fuel cap vent is not blocked. This creates a vacuum in the fuel pipe and stops the fuel flow to the carburettor.

The carburettor and primer bulb

There are many different types of small engine carburettors now on the market, including float bowl, diaphragm, pulse jet, and many more.

Remove the filter and check if there is any grass or debris build-up stopping any of the springs or governor carburettor linkage from moving. There can often be a small leaf or twig caught under the filter especially if you have been mowing under bushes. Be careful not to bend or damage any of the springs or linkage, as they are quite delicate and should be at a pre-set tension for the engine to run smoothly.

Briggs & Stratton carburettor, plan view with air filter removed, showing governor linkage.

Briggs & Stratton diaphragm and gasket.

Once the filter is removed, on Briggs & Stratton Classic, Sprint, Quattro and Quantum engines push in the primer bulb; you should see a jet of fuel squirt into the engine. If there is no jet of fuel or if the bulb returns sluggishly back to shape after depressing, it may have a blocked fuel pick up tube or jet. Look inside the tank: is there any debris, grass or water globules floating around? If so the carburettor and tank may need to be stripped and cleaned. If this is the case, take the mower or just the carburettor and tank to your local service dealer. (On many Briggs & Stratton engines, if the carburettor is mounted above the tank, the carburettor and tank will come off in one piece.) There are many good carburettor cleaners on the market, many in a spray form if you want to tackle and clean the carburettor yourself. However, if there are some jets that the carburettor cleaner won't budge, it is easier and quicker to take it to a professional. Most good service dealers have a machine called an 'ultrasonic bath', specially designed for cleaning carburettors thoroughly, including all the very fine jets that may be gummed up with fuel varnish and that are extremely hard to get at even with a high pressure air line.

On some Tecumseh engines fitted on Atco Balmoral, Suffolk Punch, many rotary mowers, cultivators and aerators, the primer bulb pumps fuel via a vacuum and it can be hard to detect if the bulb is priming properly as there is no pressure felt. There are two types – one with a vent hole and one without. On these models, try pressing the primer bulb slowly, and wait a second before releasing. Also on the vacuum primer type bulb, if the engine is hard to start try pumping the primer a few more times, nine or ten, instead of the normal six. Check if there is any fuel leaking around the primer bulb. The bulb and surrounding area should be completely dry. If you are not sure, smell your fingers: if you detect any petrol on them, either fuel or air are leaking, thus making the engine hard to start.

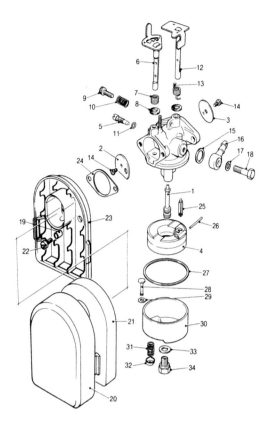

Float type carburettor showing internal parts, exploded view.

Mice love eating primer bulbs – they are not fussy especially if you leave your mower in the mouse's house (often called the garden shed or garage). If your primer bulb has a roughly shaped hole rather than a smooth split in it, it has provided a tasty treat for your resident mouse. Primer bulbs are made of a special neoprene material, which will not rot and is impervious to petrol and oil. Amazingly it is rare for mice to eat any other part of the engine or mower except for the occasional battery terminal cover and the odd one or two fuel pipes. It's easy and inexpensive to replace the primer bulb by yourself, or your local service dealer can do this for you.

To replace the primer bulb on Briggs & Stratton Sprint, Classic, Quattro and Quantum popular engines with float bowl or diaphragm carburettors:

1. Remove the air filter for better access.
2. Depress the two plastic lugs on each side of the primer ring with a screwdriver.
3. Pull the primer bulb out.

4. Refit the new replacement primer bulb, then ensure the bulb ring is located in the two side lugs, preferably with a new ring.

To make the job easier, there is a dedicated primer bulb removing tool for Briggs & Stratton engines (part no. 19461).

Good condition Briggs & Stratton primer bulb.

Mouse nibbled primer bulb; these can quite often be eaten all the way through.

ENGINE PRE-STARTING CHECKLIST

1. Check the oil level.
2. Check the fuel in the tank; ensure it's at least half full.
3. Check that the fuel is fresh.
4. Check the fuel tap is switched on.
5. Check the engine on-off switch is on.
6. Check that the primer bulb returns to shape when depressed.
7. Check the primer bulb is not leaking.
8. Check the operator presence control is fully depressed.
9. Check the choke is fully on.
10. Check the throttle lever is in start position.
11. Check the spark plug lead and cap are located securely.
12. Check the blade is not blocked with debris.
13. Check the blade(s) are not loose.
14. Check the clutch is out of drive.
15. Check the engine is on the compression stroke (when the rope has resistance).
16. Check the primer bulb for fuel or air leak (replace if any sign of a leak).

CHAPTER TWELVE

LAWN CARE

Grass is an amazing plant: for instance, having to cope with 10,000 people standing on it during a Saturday night pop concert, then with twenty-two people playing football all over it with spikes on their feet the next day!

At the start of the season, don't cut the grass too short. Give it a slow-release fertilizer as snow and rain wash out nutrients. Riding bikes and playing football on it should be avoided until it is growing and has dried. If you don't want a bowl-ing green and you are thinking of a more natural garden, try introducing some other grasses. Wild flowers bring insects, which bring mammals and birds to the garden.

Always cut when the grass is dry, whatever type of mower you use. Wet grass will stick and clog the machine; it also weighs more, putting extra strain on all parts of the machine.

Turf provides the quickest way of producing a lawn, but unless it has been specially grown from

LAWN CARE TIPS

- Mowing: April to October, twice per week.
- Mower service: December to February. (The ideal time for the lawnmower service is any time when the machine is not being used during the winter. Don't leave it till the first warm weekend of the year as this is the week when everybody gets their mowers out and when a problem arises takes it to their inundated service dealer whose workshops are lined with mower service work.)
- Feeding: starved lawns can become covered in bare patches, weeds and moss. A regular lawn feed will maintain a dense deep green sward. The best time to apply lawn food is late March to early April. This can be repeated in June or July. In October, an autumn lawn food can be applied to encourage root growth; this helps grass survive the winter and its diseases. Most people feed the plants and borders but often neglect the lawn.
- Sowing seed: late March to April or late August to September.
- Turfing: March to May or late August to September.
- Edging: late April to September.
- Scarifying: late March to April or September to October.
- Aerating: late March to April or October to November. Aerating allows air and water to reach the roots, especially if the ground is compacted. Aerating encourages grass growth and discourages the moss.
- Weed and feed: April to September.
- Lawn repair: March to September.
- Autumn feed: September to November.
- Water: when needed!

seed it may contain undesirable weeds. Seed provides the cheapest way to create a lawn; you can also choose the best mixture to suit the conditions. A fine mixture is required for close cutting lawns, picnics and relaxing; meadow grass is ideal for shady lawns. Coarse rye grass is ideal for hard-wearing lawns, football pitches and rough areas. On new lawns, cut the grass to 1½ to 2 inches. Gradually lower the cut to your desired height.

Grass grows from the root, and not from the tip like some other plants; this allows you to keep cutting it over and over again and it will always grow back. Cylinder-type lawnmowers give a much cleaner scissor cut and this allows the grass to recover more quickly; a rotary blade can thrash and rip the grass tips.

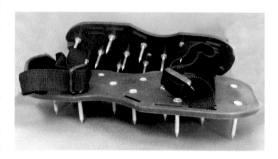

ABOVE: **Foot-powered aerator shoes: a novel idea but it can be hard work depending on soil type. Clay soil will be hard; peat and sandy ground will be softer.**

LEFT: **Domestic petrol aerator. Sharp slitting tines slice through the soil and split the grass roots, promoting water drainage, more grass growth and allowing air in for a healthy lawn.**

The petrol aerator slitting tines by Mountfield will aerate and scarify. It also features a microset depth adjustment.

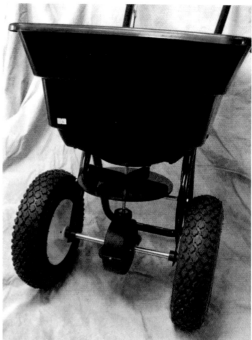

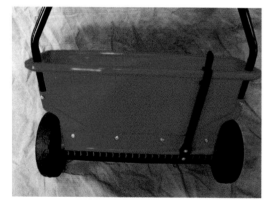

ABOVE: Rotary spreader. Its main advantage is that it will give up to a 6-foot spread.

BOTTOM LEFT: Hollow tine aerator by Sheen. The advantage of this tool is that it does not compact the soil around the hole, as it removes a plug of soil and collects it in the tray. It allows air to the grass roots – or you can fill the hole with sand for drainage.

The advantage of a drop spreader by AL-KO is that it drops fertilizer, seed, and weed control exactly where you want it.

BOTTOM RIGHT: Petrol and electric AL-KO Combi with scarifier and aerator cassette system.

CHAPTER THIRTEEN

PROFESSIONAL SERVICE AND REPAIR

Repair work

So that the service department can decipher and detect what the exact fault is on the machine, it is important to give the correct information and fault symptoms when booking it in; this can then be conveyed to the engineer when it arrives on the workbench. The following are two very common breakdown examples.

Broken recoil rope

This could be on any petrol product, including mowers, chainsaws, strimmers, hedge cutters, cultivators, generators, etc. and is a simple fault to see. When booking the machine in, ask, 'Could you fit a new rope?' In most cases it can be repaired or replaced quickly and easily, depending on the engine model. Most recoil units unscrew off very easily with either a few screws or bolts, but on some recoils (especially certain grass-trimmers, hedge trimmers and chainsaws) half the engine may need dismantling, and a degree in engineering is necessary even to attempt it. If the recoil unit is the easy type to take off, remove the recoil and just take it to the dealer. This method has the added benefit of not having to transport the whole machine; it will be repaired much more quickly and cheaply and the service agent may even have a service exchange unit to give you whilst you wait. It's worth noting down the engine make, model, and serial number in case the unit requires any spare parts

besides the rope, such as the recoil pulley, recoil pawls or recoil springs.

If, after the repair, the recoil unit or rope breaks down again within a very short time, it could be for one of a number of reasons (all recoil ropes are made of nylon, one of the world's strongest materials – often stronger than metal).

The most common causes are:

A frustrated man trying to start his mower! (Available as a birthday card from Rainbow Cards.)

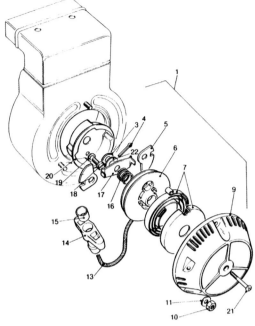

Illus. No.	Description	No. per set
1	Recoil starter assembly	1
3	Washer (1 mm thick)	as reqd
4	Split pin	1
5	Pawl	2
6	Pulley	1
7	Recoil spring and housing assembly (not supplied separately)	1
9	Recoil housing (complete with eyelet)	1
10	Nut for recoil fixing	3
11	Washer for recoil fixing	3
13	Rope	1
14	Rubber handle	1
15	Ferrule for handle	1
16	Compression spring	1
17	Activator	1
18	Screw for driving hub	2
19	Washer for driving hub	2
20	Driving hub	1
21	Recoil spindle	1
22	Washer (1.5mm thick)	as reqd

ABOVE: Atco, Qualcast, Suffolk and Webb recoil unit, exploded view.

BELOW: Atco recoil unit, exploded view. Fitted on rotary mowers.

- The engine is hard to start for some reason, and you have to pull the cord several times or have to pull it more vigorously than normal, putting premature wear and tear on the rope and the whole recoil assembly. Instead of saying the fault is the rope has snapped, the real fault is that the engine is hard to start! If you repair the rope it is going to break again in a short time if the real fault is not cured first.

- The direction and angle you pull the rope. If the centre of rope is frayed, the rope has been chafing on the recoil housing or rope guides. To remedy this, pull the rope directly out in more of a straight line, being careful to keep the rope in the centre of the rope guides as much as possible. Many models have an adjustable handle height; this alters the position of the rope. If you are a tall or short person check it's at the correct height for you as this often alters the angle of the pull. You will also

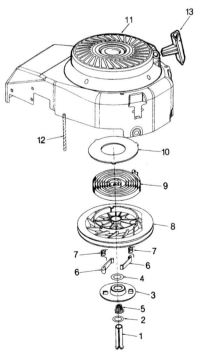

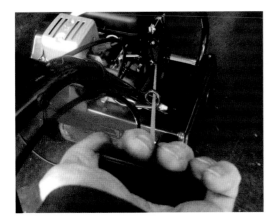

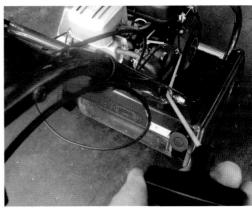

The correct way to pull the recoil rope (horizontal crank). With a strong grip on the handle, the rope is being pulled true without chafing on the rope guides, creating minimum wear and the least friction when pulled.

The wrong way to pull a recoil rope. A bad grip will often cause the recoil handle to fly out of your grip; this can cause the rope to come off the recoil pulley, which can only remedied by removing and rewinding the recoil.

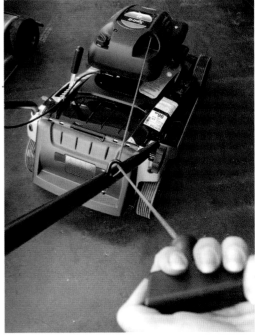

The correct way to pull a rotary recoil rope (vertical crank). Ensure the rope does not chafe on the guides. On many rotaries the rope is pulled underneath the handle, as this can make it easier to pull over.

The wrong way to pull a rotary recoil rope: pulling upward on a vertical crank engine makes more friction and is harder to pull over, causing premature recoil and rope wear.

find it is easier to pull over the engine as this causes less friction.

- Starting the engine on the wrong engine stroke. The following procedure can make the engine start much more easily, and is especially helpful on a poor starting engine and for people with perhaps a weak grip, or bad shoulder, or for those who are just not able to pull the rope quite quickly enough. It puts less strain on you and the recoil; you get maximum revolutions of the crankshaft so the engine has a much better chance of ignition (especially important when the engine is cold). Each time the engine recoil is pulled:

1. If fitted, depress the safety blade brake lever.
2. Pull the rope slowly till you feel a resistance, then stop. (Note: the resistance position of the rope can vary each time.)
3. Whilst still holding the recoil handle, let the rope rewind back slowly.
4. Pull the rope out slowly till the recoil engages with engine.
5. Whilst the recoil is engaged, grip the handle tightly, and then pull the rope briskly. At this position, the least strain on the rope and recoil assembly is achieved.
6. If the engine still does not start straight away repeat the above.

Note: when choosing a new petrol engine product, take a look where the recoil rope assembly is situated. Does it look as though it is easy to get at, compared to another one? Is it fitted on the outside of the engine cowling with the mounting bolts accessible or is it riveted? Does it look as though you will need special tools? On petrol trimmers, check where the rope comes out: if it's near the centre of the engine rather than near the side, it may require a lot more work to replace the recoil rope and should probably be done by a professional.

Outside recoil assembly.

Inside recoil assembly, a more involved repair.

Machine won't start

If on an electric machine, when the start switch is pressed the machine does not work, quite often a customer might say to the repairer, 'Please can you fit a new switch, as it's not switching on?' When the repairer fits the new switch, according to your instructions, the machine might still not work and you may end up paying for something you do not need. Unless you are completely sure the switch is faulty, do not book it in for that fault, as there may be several causes. It would be far better to say, 'Can you repair my machine? It won't start.'

Note: if you find the handles, nut and bolts are coming loose or are missing, it's a sure sign the machine is vibrating for some reason. So don't just retighten the loose bolts: the cause should be found, then the bolts retightened.

Repair options

The servicing dealer may offer you different repair options; these are just a few:

1. Full service. Includes all aspects of the machine, including cutter, bearings, engine or motor, transmission, electrics and wiring.
2. Engine/motor full service. Includes all aspects of the engine or motor – carburettor, electrics, magneto, valves checked, decoke.
3. Cutter regrind and set/re-sharpening/regrind and balance. Cutter bearings check.
4. Minor repair. A specific repair to one or two faults, for example: attention to a clutch; attention to uneven cut; attention to recoil; attention to no-start, etc.
5. Estimate. This can include anything from a full restoration service to a minor repair, letting you know exactly how much the repair is going to cost. (There may be a small inspection or diagnosis charge for this, but normally the cost is deducted if the job is accepted.) Ask for this in order to avoid receiving a huge unexpected bill. If you require the machine to be repaired quickly, it is worth asking for the basic price for the job. If the price is acceptable and as long as the machine does not require any major or costly parts or does not go above an agreed prefixed price, instruct the dealer to carry on with the repair; this way the service dealer can proceed with the job whilst the machine is stripped on the workbench, eliminating the need for putting the job on hold whilst authorization from the owner is sought.

When booking your machine in for service, ask the dealer if VAT is included in the price.

In most cases the service engineer will not know exactly which parts are required until the machine is inspected or dismantled, so all spare parts fitted would normally be extra. Most service agents will quote you a basic service price before the job is started plus any parts it may require.

Note: genuine spares are recommended; they are usually made to a higher standard and the quality is not compromised. If you do not know the make, part number, or model number of your machine, take the part you require to the servicing dealer where, if they recognize it, it can be identified for a match, or a replacement can be ordered.

Servicing your machine

There can easily be more time spent on servicing a lawnmower than on a car in many circumstances. The list opposite does not mention any seized nuts and bolts, bent or damaged blades, fractured welds, plus many more problems that may come to light during a lawnmower service.

A normal service should be carried out every twelve months depending on how many hours the machine has been used. The advantages of giving the machine a service early in the season are: peace of mind that the machine will be ready for the season's first cut; it shouldn't let you down throughout the season; it keeps the guarantee up on machines with extended warranties. The best time to have a full service or the machine repaired is when you are not using it, ideally between November and February during the winter. This is when the service dealer has the most time to service and source any obscure spares the machine may require. Regular servicing ensures optimum performance and reliability. If left later than March (depending on the weather), the servicing dealers may not be able to guarantee that your

---

FULL SERVICE

Typical full service on petrol and electric mowers would include the following:

1. Engine or motor cowls and panels removed and cleaned.
2. Spark plug cleaned or replaced. Mains plug and cables inspected and tested.
3. Cylinder head: carbon deposits removed where necessary. Cylinder bore, crankshaft, valves, compression checked.
4. Flywheel removed, magneto points refaced or replaced as necessary.
5. Electronic ignition and any electric parts checked.
6. Carburettor cleaned in an ultrasonic bath if necessary; all jets readjusted and tuned.
7. Engine oil drained and replaced.
8. Air filters and fuel filters cleaned or replaced.
9. Recoil, starter motor mechanism checked.
10. Engine revs governor operation reset and checked.
11. Motor cleaned, amps and power checked.
12. Fuel tank, pipes and cables inspected.
13. Gaskets replaced where necessary.
14. Transmission clutch and drive operation checked and tested.
15. Gears, sprockets, pulleys, chains and belts inspected and adjusted.
16. Cutter mechanism reground, reset and balanced.
17. All bearings checked and lubricated where necessary.
18. Manufacturer's electrical safety flash test carried out and logged.
19. All moving parts and oiling points lubricated.
20. All nuts and bolts, clips, screws, etc. checked.
21. Machine run and tested.

---

machine will be ready for the growing season. Most service dealers have a turnaround of seven to fourteen days, but at the height of the season this can extend well beyond this, for various reasons. At the first spell of warm weather, everyone gets their mower out for the first time and may have a problem for one reason or another. Then the machines are brought all at the same time to the dealer. Or perhaps the manufacturer may have a shortage of a spare part, or it may be obsolete or not be easy to source (especially on older machines where the manufacturer might no longer exist). The dealer may have to source parts from a specialist or have it made.

Spare parts

Some machine spares may be hard to source, especially if the machine has been originally acquired via the Internet or from a supermarket without a back-up service. Some of these machines may be imported and parts may not be available or may only be obtainable from the importer. Many machines have been re-badged, so the name on the machine may not be the name of the manufacturer. In these instances, take the machine to your local dealer for them to identify the manufacturer so parts can be sourced.

When asking for or ordering spare parts,

Briggs & Stratton serial number, located above the silencer.

Briggs & Stratton serial number positioned above the spark plug.

always give the service dealer as much information as possible regarding the make, model and serial number. Remember: not all mower manufacturers make petrol engines – they only make the chassis – so you need the make, model and serial number of the engine as well (there are many different engine manufacturers). Not all engine manufacturers make the carburettor, so on some engines you need the make and model of the carburettor. If this information is not available, the dealer needs to use his detective skills to decipher exactly what machine you have before ordering any parts.

If you were the owner of a Suffolk Punch, for example, here are some simple questions that you might be asked by a service department in order to identify your specific model:

- Is it petrol or electric?
- What size is it (12in, 14in, 17in or 18in)?
- Is it a cast-iron, aluminium or steel chassis?
- Is it a five-, six-, or ten-blade model?
- Is the engine 75cc or 98cc?
- Is the engine a Suffolk, Qualcast, Tecumseh, or Kawasaki?
- Is it a metal or plastic recoil model?
- Is it a Qualcast or Tecumseh key start model?

- Is it a QX model?
- Does it have a pram handle?
- Does it have points or electronic ignition?
- How many fins on the flywheel?
- Is it a Super Suffolk Punch, a Suffolk Punch Professional or a Suffolk Punch MK7?
- Is it a Dual Drive?
- Is it a four-wheel or rear-roller rotary; push or self-propelled, chain or belt drive?
- Is it a red or green model?
- How old is it?
- Has it got a dipstick?
- How many controls on the handlebar?

Clearly, having the correct information can save you a lot of time and confusion – even more so on some makes, as a spare part is almost impossible to order without a correct part number, model or serial number.

When ordering some parts for cylinder lawnmowers, one part can be mistaken for another. For example, many lawnmowers have several clutches (a clutch for the drive, a clutch for the engine and a clutch for the cutter), and they can easily be confused. If you do not know the part number, ensure that the parts department understands which clutch it is.

CHAPTER FOURTEEN

LAWNMOWER RACING

The first ride-on lawnmower was made in 1904, manufactured by Ransomes. The new contraption was bought by Cadbury Chocolates at Bournville House for their workers' sports field. Little did they know that within seventy years people would be racing ride-on mowers all over the world.

Lawnmower racing is a unique and great British motor sport, inexpensive, and great fun. Racing lawnmowers compete all over the UK, USA and Europe. Speeds of 60mph are often attained; anything above this and racing lawnmowers can seem to have a mind of their own – after all, the original lawnmower design had a maximum speed of only 6mph.

It all started in 1973 at the Cricketers Arms pub in Wisborough Green, West Sussex, where people were talking about racing cars. The conversation went from rally cars to bar stool racing, to racing combine harvesters, but unfortunately there were only three in the county and the farmers were reluctant to lend them – let alone race

RIGHT: **Class II prepared racing lawnmower.**

BELOW RIGHT: **Ludlow 3: a racing lawnmower on full throttle can attain speeds of over 60mph. Note the lowered seat and handlebars – this gives it a much lower centre of gravity.**

BELOW LEFT: **A racing lawnmower about to turn a traffic cone into grass clippings.**

them. Eventually one person said, 'Everyone's got a lawnmower – let's race them!' They attracted an entry of thirty-five drivers with mowers ranging from a 1923 Atco to a brand new 8hp tractor and so the British Lawnmower Racing Association was born.

There were races for run-behind mowers, towed-seat mowers and the type you sit on. The run-behind class is governed merely by the ability of the perspiring runner to stay with the machine. The sit-on class originally was the slowest, attaining 7mph, but over the years with attention to gearing and centre of gravity it has become the fastest. Once the roller seat and handlebars have been lowered, the driver has much more control. It's a soul-stirring sight to watch a field-full of these machines, grassbox to grassbox, exhausts bellowing, round a tight circuit at 50mph.

Sir Stirling Moss, the famous Formula One champion, was attracted by the club atmosphere and fun of racing and held the championship in 1975/6, exclaiming, 'It's just such fantastic great fun!' Over the years the sport has grown rapidly. By the end of the 1970s over a hundred enthusiasts had their own racing mowers in this phenomenally successful sport. There is now an annual twelve-hour race with a Le Mans-type start held at Wisborough Green. It starts at 10pm on the Saturday closest to midsummer's day and ends twelve hours later at 10am on the Sunday. After a night of thrills and spills, the original race was won by famous racing drivers Stirling Moss, Derek Bell and Tony Hazelwood. By 1980 Derek Bell wrote himself into the Guinness Book of Records by covering 276 miles in a lawnmower race.

Chris Tarrant, Noel Edmonds, Mike Smith, and many more have all been keen on lawnmower racing. During the twelve-hour night racing Oliver Reid allegedly ran someone over in the ladies' toilet tent; everyone said he was half cut! A screaming lady was observed, who sped like a

banshee into the night and to this day has never been identified.

Other big events on the national calendar are the two-day World Championships. This event has attracted entries from as far afield as Zimbabwe and New Zealand, with the superb marble trophy being captured by the Hong Kong entry.

Lawnmower racing comes under strict rules, just like Formula One racing; anybody not con-

The 'Crossley Flyer', a Class II racing lawnmower, originally a scrapped council Hayter Condor and renamed in 1980. It was a cut above the rest, went on to win many trophies in the eighties and was featured on many TV programmes. But it could be a bit of a 'sod' to handle at speed.

The 'grass roots' of lawnmower racing, the annual six-hour and twelve-hour races.

ABOVE LEFT: **Wray 2: the common sight of a Class III racing mower in full flight.**

ABOVE RIGHT: **Don't let the grass grow under your feet! Racing at Barnard Castle on a 1970s Westwood Lawnbug.**

BELOW RIGHT: **Scorton, cutting a dash!**

forming is 'turfed out'. Crash helmets and protective race suits are always adorned with a special engine cut out lance attached to the driver, should machine and driver become detached. Here are the grass roots of the regulations:

1. The machine must originally have been bought to cut grass, and not the rolling prairies of America.
2. You can't fit a larger engine than the manufacturer originally intended.
3. It must still look like a lawnmower.
4. The blades must be removed.

There are three classes:

1. Class I: Ride-on, garden tractor type.
2. Class II: Sit behind, cylinder lawnmower.
3. Class III: Run behind, a walk-behind machine made to go a bit quicker than you can run. This class of machine is often raced in teams.

A CLASS III RACING MOWER

Picture Brian Radam on Noel Edmonds' *Late, Late Breakfast Show* with his Class III racing machine, originally a 1978 Westwood Lawnbug, slightly modified with a 320cc Briggs mower engine, with bits from Jaguar steering, Vauxhall suspension and a canteen seat. The lawnmower racing heats were staged on an airstrip along with the world's fastest jet-powered lorry. It started when Noel Edmonds announced on prime time BBC TV: 'Brian Radam always wanted to do it on the grass in front of the viewing public.'

Brian's wife Su raced this machine and once, whilst dicing at high speed with other drivers, ended with a spectacular racing incident: she rolled the machine several times and was turfed out of the seat and thrown through the air. Racing marshals rushed to the scene; fortunately for Su, not much damage was inflicted on the mower, which went on to win the Championship.

The author with his championship-winning Class III racing mower.

An old scrap ride-on mower that's seen better cutting days is ideal and is the grass roots of the sport, the essence being that it will not cost an arm or a leg (unless you run over yourself), as they can be bought for just a few pounds. Most original engines are plenty powerful enough to race as the average larger mower or tractor has an adequate 300cc or bigger engine fitted. The main modification is to alter the gear ratios, lower the centre of gravity and remove the blades.

Mower racing could be the answer for many would-be motor sport enthusiasts who have been put off by the horrendous cost, especially if the old mower in the shed can't be fixed. We are not saying it's a perfect substitute, but, having once sat behind a screaming, bucking and almost out of control lawnmower at speeds of over 50mph while in close contact with a dozen similar madmen and machines roaring around a bumpy track in a stubble field, you have to admit it's an awfully good alternative! (Full, in-depth lawnmower racing plans on 'How To Make Your Own Racing Lawnmower' can be obtained from the British Lawnmower Racing Association.)

The world lawnmower land speed record of 87.833mph was achieved on 23 May 2010, driven by Don Wales, grandson of Sir Malcolm Campbell at Pendine Sands, Wales, the scene of the first land speed record in 1926 reaching 146mph. The racing lawnmower is also capable of cutting grass.

For more information about lawnmower racing, see Further Information (page 106).

CHAPTER FIFTEEN

LAWNMOWER SAFETY

- Always read the instruction manual supplied with your machine.
- Clear the area of stones, objects and debris before mowing.
- Never leave a mower unattended with the engine running, especially if there are children nearby.
- Never mow barefoot. Wear strong shoes or boots and trousers.
- Keep hands and feet well away from the blades.
- If the machine starts to vibrate, stop and investigate.
- If the machine starts making a strange noise, stop and investigate.
- Use a circuit breaker on mains electric machines, such as an RCD (Residual Current Device). This safety product ensures you will not get an electric shock if anything untoward happens to the mains cable or machine. They are inexpensive and can be used on any household electric appliance, simply fitted in seconds. It is an adaptor that just plugs into any 13amp mains socket.
- Prevent accidental starting: always remove the spark plug wire when servicing.
- Never use a mains electric mower in the rain or when the grass is wet.
- Take care to avoid running over the mains cable.
- Disconnect the mains plug before making any adjustments on electric machines.
- Check the mains cables for any cuts or frays. Replace the cable if damaged.
- Regularly check that all securing nuts and bolts are tight.
- Replace damaged or missing items immediately.
- Ensure blades are not worn, damaged, bent or cracked.
- Do not use an unbalanced cutter blade.
- Do not run an engine in an enclosed area.
- Always turn the engine off before filling with fuel.
- Do not re-fuel indoors.
- Store petrol in a cool and dry place.
- Do not store petrol near an open flame, pilot light or spark.
- Wipe up any fuel spills or let fuel evaporate.
- Move machine away from any fuel spillage before starting.
- Avoid transporting machine with fuel in the tank.
- Replace all fuel container caps securely.
- Check fuel lines etc. for any leaks.
- Do not overfill fuel tank – allow space in tank for expansion.
- Empty the fuel by running the engine dry, if tipping or turning the machine on its side.
- Check which way the engine can be tipped if oil is in the sump.
- Beware: young children are very inquisitive; sometimes they like feeding lawnmower petrol tanks with water or soil. Store and lock the machine away securely, especially if children are about.

APPENDIX: MONTHLY LAWNMOWING GUIDE

January/February

If the weather is mild, the grass can occasionally be cut, but keep it above 1in. Do not walk on the grass if it is waterlogged or frozen.

March

Aerate and rake the lawn. This will encourage the grass and discourage the moss.

April

Apply a fertilizer and moss killer if required. Carry on mowing on a slightly lower height of cut.

May

Apply a weed and feed or a selective weed killer if the lawn was not fed in April. Carry on mowing.

June

Mow as often as possible; the secret is 'little and often' – twice a week is ideal. Water if necessary.

July

Carry on mowing, fertilize and water if necessary. If hot, don't cut too close.

August

Carry on mowing and watering as necessary. If cracks appear in drought conditions, a soil and sharp sand can be used to fill in.

September/October

Slightly raise the height of cut. Apply autumn/winter fertilizer. Apply weed and moss killer if necessary.

November/December

Do not mow if the grass is wet as this will compact the ground and encourage moss and waterlogging. Use a stiff brush to remove any wormcasts.

GLOSSARY

Some of the following garden tool and machinery-related words may have different definitions in other applications; these refer only to garden machinery.

ABS Acrylonitrile butadiene styrene, a strong copolymere plastic-type material used in mower chassis.

AC Alternating current, as from a mains supply.

Actuator Device in a strimmer that allows the nylon line to feed out.

Aerator shoes Pair of clamp-on soles with spikes to aerate the lawn.

Air gap Precise gap measurement between a coil and flywheel magnet.

Air vane Device worked by air flow to govern an engine speed for smoothness.

Alternator Electrical device that generates alternating current.

Ampere (amp) Unit for measuring electrical current.

Antisyphon Valve inserted in a line to prevent backflow.

Anvil Used in garden shears: a flat blade against which a sharp blade is pressed.

Armature The copper wire windings on the rotating shaft of a brush motor. Also describes an electronic coil on a petrol engine.

Backlap To sharpen cylinder blades, applying grinding paste and turning the blades backwards.

Banjo A banjo-shaped carburettor fitting, to which the fuel pipe is attached.

BBC Blade brake clutch. Device for stopping the blades without stopping the engine.

Belt drive V-belt or tooth belt that transfers power from a power source to blades, wheels, roller or drive.

Big end The part of the connecting rod which fits around the crankshaft.

Blade boss Adaptor, which fixes the blade to a drive shaft.

Blower housing Engine cover over the flywheel. American term for an engine cowling.

Bottom blade The fixed blade on a cylinder lawnmower.

Bowden cable A flexible control cable made of thin wire strands.

Broadcast To spread fertilizer, weed killer, seed, etc.

Broadcast spreader Rotary spreader that throws and spreads granulated product.

Brush Carbon conductor found in electric motors, which presses against the commutator.

Bump feed Spring-loaded cup fitted on the head of a grass trimmer, allowing the nylon line to be fed out when tapped on the ground.

Cam Pear-shaped projection on a shaft which moves an adjacent part.

Camshaft Rotating shaft with cams attached.

Capacitor Electrical component that stores electrical energy, used on an electric motor.

Chain brake Chainsaw device that stops the rotation of the saw chain.

Chain drive Bicycle-type chain that transfers a power source to the blades, wheels or roller.

Chain skate Nylon or metal slipper for the smooth running of a chain, normally fitted on the chain adjuster.

Circlip Sprung steel round clip that can be widened to fit in a slot on a shaft.

Clam shell Housing (usually in pairs: left- and right-hand), which covers and mounts the motor on an electric trimmer.

Clearance The space between two moving parts, or a moving and a stationary part.

Clutch Device connected to a drive shaft which can be engaged or disengaged.

Coil Wire wrapped around a core, providing voltage to the spark plug.

Commutator A cylinder of insulated brass bars on the rotating shaft of a motor against which the brushes collect current.

Con rod Connecting rod, joining the crankshaft to the piston.

Condenser Electrical component that stores electrical energy, used on a petrol engine magneto.

Cotter pin Pin with a split end; when put through a hole can be spread apart, to hold parts together.

Cow-horn handle Two-handed, U-shaped handle often fitted on brush cutters.

Crank Short for crankshaft.

Crankshaft Rotating shaft driven by a piston, which delivers power to the gears or drive.

Cup and cone Open-type bearing race which runs on a cone-shaped shaft and can be adjusted by tensioning the bearing onto the cone.

Cutter bar The blade on a rotary mower in the UK. The bottom blade of a cylinder mower in the USA.

DC Direct current, power from a battery.

Decal Sticker or transfer depicting a manufacturer, instructions or diagram.

Deck The main part of a rotary chassis.

Decompressor Device on an engine to lower the compression to make it easier to pull over.

Diaphragm Flexible oscillating membrane, as found in carburettor.

Dipstick Measuring device which indicates the level of oil. (Also a slang term in the motor and engineering industry for a 'wally'. Often found on the end of a telephone.)

Direct collect Term used for a garden tractor with a system where the grass is thrown directly into the grassbox.

Double insulated In electrical machines, designed not to require a safety connection to electrical earth. Designed to prevent electric shock to the user.

Drop spreader Tool for dropping seed or fertilizer to the exact width of the machine.

E clip A sprung steel round clip, similar to the shape of an E, which fits in a slot on the outside of a shaft.

Electrode A conductor, such as the tips on a spark plug.

Electrolyte The liquid in a lead acid battery, consisting of sulphuric acid and distilled water.

Electronic ignition Solid state component coil that requires no points or condenser.

Element A paper type air filter.

Engine knock Noise made by an engine if there is too much carbon in the combustion chamber, if the flywheel is loose, if the blade boss is loose, if the big end is worn, or the engine is running on a poor grade of fuel.

ES Electric start or button start engine.

Feeler gauge Metal strips used to measure narrow gap dimensions, as the gap on spark plugs, points electrodes and electronic coil air gaps.

Ferule Metal cup that secures the outer sleeve on a control cable. The outer part on a tool that secures the handle (as on a pair of shears).

Flexible drive Rope-like flexible wire that transmits power in a curved shaft like a bent shaft grass trimmer.

Float Floating device, which stops or allows the flow of fuel in a carburettor.

Float bowl The bowl-like container for the float.

Flooded Where the carburettor or engine have been overfilled with fuel.

Flywheel key Metal key that connects the flywheel to the crankshaft; if the flywheel is strained the key will break, protecting the flywheel or shaft from damage.

Four-stroke engine An engine with four strokes to one complete power cycle: intake stroke, compression stroke, power stroke and exhaust stroke

Friction disc Device to let a blade slip when under impact.

Galvanized Zinc coating on steel or iron to prevent rust.

Gear Round transmission drive part with teeth contacted and driven by another gear.

Governor Device coupled with the throttle to limit the engine speed, and stop an engine from over-revving. Aids smooth running.

Governor spring Spring controlling the engine speed and smoothness.

Grease nipple A fitting shaped like a nipple, allowing grease to be pumped one way into a machine part such as a bearing or shaft.

Grind and set Term for resharpening and readjusting the set on a cutting cylinder of a lawnmower.

Grinding compound/paste Paste made for grinding in blades or valves.

Gudgeon pin Pin that connects the small end of the con rod to the piston.

Gum Varnish-like substance left in an engine when fuel goes stale.

Hand wheel Nut and bolt, where the nut can be undone or tightened by hand.

Hollow tine aerator Tool that takes a plug of soil from the ground, does not compound the soil around it.

hp Horsepower, a measurement of power of an engine. Ransomes Lawnmowers say they invented this term.

HT lead High tension lead. This wire, conveying voltage, connects the spark plug to the coil.

Hydro/hydrostatic Hydraulic variable speed gearbox.

Hydrometer Device that checks the condition of a lead acid battery by measuring the specific gravity of the electrolyte.

Idler pulley Moving pulley that tensions a drive belt.

Impeller Disc with fins, creating air flow to cool motors. Used on all hover-type mowers to create the lift.

Induction motor Brushless motor, requiring less maintenance than a brush motor.

Interlock switch Safety switch in an electrical circuit which prevents operation unless a component is properly engaged, often found on garden tractors.

Keyway Slot in a shaft, which houses a locating key to stop a pulley or sprocket from turning.

KS Key start engine.

Lawnmowerist A new word coined to describe one who is interested in, or who collects, lawnmowers.

LCD Liquid Crystal Display. An instruction or information screen.

Lead acid battery Battery containing lead, sulphuric acid and distilled water. Often used in tractors and cars.

Lithium ion High-energy battery, lighter and more powerful than others, with no memory effect.

Little end The part of the con rod which fits around the gudgeon pin in the piston.

Magneto Electrical components that convert mechanical energy to electrical energy. Mainly the electric parts that produce the spark on a mower engine.

Mills pin Solid pin, wider in the centre to grip when fitted into a shaft.

Motor brake Magnetic device fitted to an electric motor, which stops the motor as soon as a switch is released.

Muffler American word for an engine silencer or exhaust.

Mulching Term for recycling grass clippings

by finely cutting and re-cutting the grass, concealing the cuttings within the standing grass to wilt, decompose and act as fertilizer.

Nickel cadmium Rechargeable battery that uses nickel oxide hydroxide and metallic cadmium as electrodes.

NYLOC A nut with a nylon insert to prevent it coming loose.

OHC Overhead cam: a camshaft that drives the valves above the piston.

OHV Overhead valve: engine valves positioned on top of the piston, promoting more efficient running than a side valve engine.

Oil seal A rubber ring, often with a coil spring inserted, to fit on a shaft next to a bearing to prevent oil seepage.

Oil splasher/oil dipper A bar fitted onto the big end to distribute oil around the engine.

Oiler A fitting with a hinged cap, to keep debris out, allowing access for oil to lubricate parts like bearings and shafts.

Oilite bearing Bearing made of phosphor bronze with microscopic holes, which hold lubricant within itself.

OPC Operator Presence Control: a handle grip which when released will stop the machine. Often called a 'Dead Man's Handle'.

'O' ring Rubber or neoprene sealing ring.

Pawl A part which when moved engages another part, as in a recoil or gears.

PD Power drive or self-propelled model.

PDI Pre-delivery inspection. Model is fully assembled, oil and fuel fitted, run, tested and ready to use, then often demonstrated to the customer.

pH The relative acidity or alkalinity of soil.

Pinion Small, round gear that normally engages a larger gear.

Pitted Depressions in metal parts caused by corrosion, rust or improper combustion.

Points Contact breaker points. A pair of open and closing electrodes which control the ignition timing of the spark.

Points gap A precise measurement between the two contacts on the contact breaker points.

Powered sweeper An attachment to a tractor which sweeps the grass up by a revolving brush, driven by a PTO drive.

Primer bulb A neoprene bulb which when pushed draws fuel into the carburettor.

psi Pounds per square inch pressure.

PTO Power Take Off: a facility to attach auxiliary equipment found on garden tractors via a drive shaft or belt for a powered tool such a grass sweeper, shredder, chipper, etc.

Push-on retainer A round washer with sharp tangs to be pushed over a shaft, which cannot slip back.

Quill housing The bearing housing, often complete with bearings and shaft, which the blade attaches to on a cutter deck on a garden tractor.

RCD Residual Current Device. An inexpensive lifesaver that reduces the risk of electrocution. Monitors current in the live and neutral wires, which should be equal; when residual current is present due to a fault or short the RCD cuts the power. Test the RCD each time it is used by pressing the test button, then reset.

Reciprocating Knife blades which cross over each other, as fitted on hedge cutters, Allen-type scythes and barbers' trimmers.

Recoil Rope device fitted onto an engine crankshaft, were the rope retracts into a housing after pulling.

Rectifier Solid-state component converting alternating current (AC) to direct current (DC).

Reed valve Thin metal plate mounted between the crankcase and carburettor on a two-stroke engine.

Reel mower American name for a cylinder mower.

Regrind and balance To re-sharpen and adjust the balance on a rotary cutter blade.

Ring gear Gear that is on the inside or outside of a circumference.

Roll pin Hollow tubular pin, supplied in various sizes with a split along its length, which can be fitted tightly in a hole.

ROS Reverse Operation System. A switch device that allows a garden tractor to cut in reverse.

Roto-stop Name used by Honda to describe a mechanism that stops the blade from turning but keeps the engine running. (Also known as a blade brake clutch.)

'Roundtuit' An indispensable gift: once accomplished, all your mower service problems will magically vanish. Often echoed around the country when the mower wants fixing, 'I must get roundtuit. Tuits seem to be a rare commodity, especially the round ones!!

RPM Revolutions per minute, a measurement of the speed of a shaft.

Shanks's Pony Pony-drawn lawnmower made by Shanks in the 1800s. An old-fashioned term meaning 'walking'.

Shim A thin piece of metal to take up a gap or play in between parts.

SIF Suffolk Iron Foundry.

Solenoid Coiled wire which, when electrified, becomes an electromagnet. Often used as part of an electric start. A heavy duty switch on older battery electric models.

SP Self-propelled or power-driven model.

Spark plug gap The space between the electrodes on a spark plug.

Spherical bush A ball-shaped phosphor bronze bush, which can be rotated to compensate for any change in the angle of a shaft.

Spider Part of the cutting cylinder to which the blades are attached.

Sprocket A wheel with teeth driven by a chain.

Stator plate The mounting plate, to which the coil, condenser or points are fixed.

Strimmer Black & Decker trademark name for an electric nylon line grass trimmer.

Sub seal A seal that fits next to a bearing to stop the ingress of debris.

Suppressor cap Connects on top of the spark plug, stops radio interference.

Tachometer Device for measuring the speed of an engine.

'Tap 'n' go' Grass trimmer term, same as 'bump feed'.

Tappet In an engine, a rod that is moved by a cam, which operates a valve.

Thatch Dead grass and clippings that form a mat at the base of grass.

Tickler Spring-loaded rod, which depresses the float to allow fuel to flow into a carburettor, found on many pre-1980s lawnmowers.

Top dead centre When the piston is at the top of the engine cylinder bore.

Torque Pulling or turning power of an engine produced by rotating shaft. Many engines are now rated in torque instead of horse power.

Torque wrench Tool with a gauge indicating the torque applied to tightening a bolt, measured in foot pounds.

Towed sweeper Grass-collecting accessory with revolving brushes, powered by the forward motion of its own drive wheels. Normally pulled by any non-collecting tractor.

Trunion nut Tubular-shaped threaded nut, which can move around an axis.

Two-stroke Engine that has one power stroke for every two piston strokes.

Ultrasonic bath Device for cleaning carburettor parts.

Vapour lock Condition caused by overheating of fuel where bubbles obstruct fuel flow, causing the engine to stall.

Venturi A trumpet-shaped part of a carburettor that accelerates air flow into the engine, and creates greater velocity.

Weedwhacker American brand name for a grass trimmer.

Woodruff key Often half-moon shaped, a key that sits in a slot in a shaft to stop a pulley, gear, wheel or roller from moving.

FURTHER INFORMATION

Speciality publications

Briggs & Stratton

Service Tools Catalogue. Speciality service tools designed for repairing Briggs & Stratton petrol and diesel engines. Part no. MS-8746.

Small Engine Care & Repair Manual. 143-page, step-by-step guide with over 300 Briggs & Stratton photos. Part no. 274041.

Small Engines. 332-page textbook of Briggs & Stratton history and development, scientific principles, troubleshooting, engine applications etc. Part no. CE8020.

Major Engine Failure Analysis. 47 pages, packed with photos showing a variety of engine failures and how to analyze cause of failure. Part no. CE8034.

The Legend of Briggs & Stratton. 208 pages of historical data, and classic photos from 1908 to the present. Ideal for every enthusiast or collector. Part no. MP-5160.

Single Cylinder 'L' Head Engine Repair Manual. 350-page manual containing detailed and easy-to-follow instructions on how to adjust, tune up and repair Briggs & Stratton Engines. Part no. 270962.

Vanguard Single Cylinder OHV Air-cooled Engines. Repair manual for Vanguard single cylinder overhead valve engines. Part no. 272147.

Twin Cylinder 'L' Head Repair Manual. Covers twin cylinder Briggs & Stratton engines. Part no. 271172.

Vanguard Repair Manual. For V-Twin overhead valve engines (OHV), series 290400, 290700, 303400, 303700, 350400, 350700, 351400, 351700, 380400, 380700. Part no. 272144.

Intec Repair Manual. V-Twin (OHV) engines. Model Series 405700, 406700, 407700. Part no. 273521.

2-Cycle Single Cylinder Lawnmower Repair Manual. Briggs & Stratton. Part no. 800100.

Antique Repair Manual. Covers out-of-production Briggs & Stratton engines built 1919–1981. Part no. CE8069.

Popular Parts Source. 130 pages with hundreds of pictures. Covers Briggs & Stratton engine parts and accessories. Quick reference chart to find the right part for the right application.

Accessories Catalogue. Every accessory for Briggs & Stratton engines. Part no. MS-3880.

Tecumseh

Tecumseh Mechanic's Handbook. 109-page, comprehensive manual, covering all two- and four-stroke engines from 3hp to 11hp. Part no. MS-4185-4/05.

Tecumseh Technician's Handbook. 115 pages, covers 3hp to 11hp engines.

Lawnmower racing

How to Build your own Racing Lawnmower. Comprehensive book of rules, specification and build manual.

Magazines

Stationary Engine Magazine. 60-page monthly magazine, covering all types of vintage engines. Articles, hints and tips, helpline and advertisements for machines for sale. Kelsey publishing, Kent.

Tractor & Machinery. Over 180-page magazine of all types of tractor, articles and wide variety of related items for sale. Kelsey Publishing, Kent.

On DVD

Lawnmowerworld: The Movie. DVD film about the British Lawnmower Museum. Features the unique 'Lawnmower Song', written and sung by Doug Miles, a direct descendant of the inventor of the lawnmower. The words depict the history of the lawnmower.

Useful UK contact details

Suppliers

Alko, Wincanton Business Park, Somerset BA9 9RS. Tel: 01963 828 000. Fax: 01963 828001. Website: www.alkogarden.co.uk.

Allen, Spellbrook, Bishops Stortford, Hertfordshire CM23 4BU. Tel: 01279 723444. Fax: 01279 723821. Website: www.hayter.co.uk.

Allett, Hanger 5, New Road, Hixen, Staffordshire ST18 0PJ. Tel: 01889 271503. Fax: 01889 271321. Website: www.allett.co.uk.

Ariens, Waterloo Industrial Estate, Waterloo Road, Bidford on Avon, Warwickshire B50 4JH. Tel: 01789 490177. Fax: 01789 490170. Website: www.claymoregrass.co.uk.

Atco, PO Box 98, Broadwater Park, North Orbital Road, Denham, Uxbridge, Middlesex UB9 5HJ. Tel: 0844 736 0108. Website: www.atco.co.uk.

Atco Car Owners Club, British Lawnmower Museum, 106–114 Shakespeare Street, Southport, Merseyside PR8 5AJ. Tel: 01704 501336. Fax: 01704 500564. Website: www.lawnmowerworld.com. Email: atcocarclub@lawnmowerworld.com.

Black & Decker, 210, Bath Road, Slough, Berkshire SL1 3YD.Tel: 01753 574277. Fax: 01753 551155. Website: www.blackanddecker.com.

Bosch, Gipping Way, Stowmarket, Suffolk IP14 1EY. Tel: 01449 742031. Fax: 01449 742214. Website: www.boschgarden.co.uk.

Briggs & Stratton, Road Four, Winsford Industrial Estate, Winsford, Cheshire CW7 1SZ. Tel: 01606 868276. Fax: 01606 862180. Website: www.briggsandstratton.com. Email: service@briggsandstratton.co.uk.

Brill, Wincanton Business Park, Wincanton, Somerset BA9 9RS. Tel: 01963 828000. Fax: 01279 723821.

British Agriculture & Garden Machinery Association, First floor, Entrance B, Salamander Quay West, Harefield, Hillingdon UB9 6NZ. Tel: 0870 2052834. Fax: 0870 2052835 www.bagma.com. Email: info@bagma.com.

Champion Spark Plugs, Arrowbrook Road, Upton, Wirral, Merseyside L49 0UQ. Tel: 0151 678 2424. Fax: 0151 677 0098. Website: www.championsparkplugs.com.

Countax, Countax House, Great Haseley Trading Estate, Great Haseley, Oxfordshire OX44 7PF. Tel: 01844 278800. Fax: 01844 278792. Website: www.countax.com.

Dennis, Ashbourne Road, Kirk Langley, Derby DE6 4NJ. Tel: 01332 824777. Fax: 01332 824525. Website: www.dennisuk.com.

E.P. Barrus, Launton Road, Bicester, Oxon OX26 4UR. Tel: 01869 363636. Fax: 01869 363620. Website: www.barrus.co.uk.

Etesia, Unit 12, Hiron Way, Budbrooke Industrial Estate, Warwick SG4 7AS. Tel: 01926 403319. Fax: 01926 403323. Website: www.etesia.com.

Flymo, Oldends Lane Industrial Estate, Stonedale Road, Gloucestershire GL10 3SY. Tel: 01453 820300. Fax: 01453 826936. Website: www.husqvarna.co.uk.

Friendly Robotics, Pennells Complex, Newark Road, South Hykeham, Lincoln LN6 9NH. Tel: 0845 1235844. Website: friendlyrobotics.co.uk.

Gardena, Preston Road, Aycliffe Industrial Park, Newton Aycliffe, County Durham DL5 6UP. Tel: 0844 8444558. Fax: 01453 826936. Email: info.gardena@husqvarna.co.uk.

GGP (Global Garden Products), Bell Close, Newnham Industrial Estate, Plympton, Plymouth Pl7 4JH. Tel: 01752 231500. Fax: 01752 231645. Website: www.mountfield.co.uk.

Hayter, Spellbrook, Bishops Stortford, Hertfordshire CM23 4BU. Tel: 01279 723444. Fax: 01279 600338. Website: www.hayter.co.uk.

Honda, 470 London Road, Slough, Berkshire SL3 8QY. Tel: 0845 2008000. Website: www.honda.co.uk.

Husqvarna, Preston Road, Aycliffe Industrial Park, Newton Aycliffe, County Durham DL5 6UP. Tel: 0844 8444558. Fax: 01453 826936. Website: www.husqvarna.co.uk.

Jonsered, Preston Road, Aycliffe Industrial Park, Newton Aycliffe, County Durham DL5 6UP. Tel: 0844 8444558. Fax: 01453 826936. Website: www.husqvarna.co.uk.

Kawasaki, 1 Dukes Meadow, Millboard Road, Bourne End, Buckinghamshire SL8 5XF. Tel: 01628 856750. Fax: 01826 856796.

Kohler, Unit 1 Europark, Watling St, Clifton upon Dunmore, Rugby, Warwickshire CV23 0QA. Tel: 01788 861150. Fax: 01788 860450. Website: www.hardi-uk.com.

Kubota, Dormer Road, Thame, Oxfordshire OX9 3UN. Tel: 01844 214500. Fax: 01844 216685. Email: service@kubota.co.uk.

Lawnflight, Launton Road, Bicester, Oxfordshire OX26 4UR. Tel: 01869 363636. Fax: 01869 363620. Website: www.lawnflight.co.uk.

Lawnmower Warehouse Southport, 106–114 Shakespeare Street, Southport, Lancashire PR8 5AJ. Tel: 01704 501336. Fax: 01704 500564. Email: info@lawnmowerworld.com.

Lloyds & Co., Birds Hill, Letchworth, Hertfordshire SG6 1JE. Tel: 01462 683031. Fax: 01462 481964. Website: www.lloydsandco.com.

Makita, Michigan Drive, Tongwell, Milton Keynes, Buckinghamshire MK15 8JD. Tel: 01908 211678. Fax: 0800 454 746. Website: www.makitauk.com.

McCulloch, Preston Road, Aycliffe Industrial Park, Newton Aycliffe, County Durham DL5 6UP. Tel: 0844 8444558. Fax: 01453 826936. Website: www.husqvarna.co.uk.

Mitsubishi, Master Farm Services Ltd, Bures Park, Colne Rd, Bures, Suffolk CO8 5DJ. Tel: 01787 228450. Fax: 01787 229146.

Mower Magic, Pennells Complex, Newark Rd, South Hykeham, Lincoln LN6 9NH. Tel: 0845 1235844. Website: www.mowermagic.co.uk.

Mountfield, GGP, Unit 8, Bluewater Estate, Bell Close, Plympton, Plymouth PL7 4JH. Tel: 01752 231500. Fax: 01752 231645. Website: www.mountfield-online.co.uk.

MTD, Launton Road, Bicester, Oxfordshire OX26 4UR. Tel: 01869 363636. Fax: 01869 363620.

Murray, Road Four, Winsford Industrial Estate, Winsford, Cheshire CW7 1SZ. Tel: 01606 868276. Fax: 01606 862180. Website: www.murray.com.

NGK Spark Plugs, Maylands Avenue, Hemel Hempstead, Hertfordshire HP2 4SD. Tel: 01442 281000. Fax: 01442 281001. Website: www.ngkntk.co.uk.

North West Lawnmower Racing Association: Secretary, Tony Dwight, Tel: 07802 669376; Darren Whitehead, Tel: 07710 770303.

Old Lawnmower Club, P.O. Box 5999, Aspley Guise, Milton Keynes MK17 8HS.

Partner, Preston Road, Aycliffe Industrial Park, Newton Aycliffe, County Durham DL5 6UP. Tel: 0844 8444558. Website: www.partner.co.uk.

Qualcast, Gipping way, Stowmarket, Suffolk IP14 1EY. Tel: 01449 742031. Fax: 01449 742214. Website: boschgarden.co.uk.

Ransomes Jacobsen, West Road, Ransomes Europark, Ipswich, Suffolk IP3 9TT. Tel: 01473 270000. Website: www.jacobsen.com/europe.

Robin America Inc., Fuji Industries Ltd Group, 905 Telser Road, Lake Zurich, Il 60047n (800) 277-6246. Tel: (847) 540-7300. Website: www.robinamerica.com.

Robomow, Unit 11, Westminster Industrial Estate, Station Road, North Hykeham, Lincoln, Lincolnshire LN6 3QY. Tel: 01522 690005. Fax: 01522 690004. Website: www.robomow.eu/www.robomow.com.

Rover, Farrell House, Orchard Street, Worcester WR5 3DW. Tel: 01905 763027. Fax: 01905 354241. Website: www.midlandpower.co.uk.

Ryobi, Medina House, Fieldhouse Lane, Marlow, Bucks, BL7 1TB. Tel: 01628 894400. Fax: 01628 894401. Website: www.ryobipower.co.uk.

Shanks, Launton Road, Bicester, Oxon OX26 4UR. Tel: 01869 363636. Fax: 01869 363620. Website: www.barrus.co.uk.

Sisis Equipment Ltd, Hurdsfield, Macclesfield, Cheshire. SK10 2LZ. Tel: 01625 503030. Fax: 01625 427426. Website: www.sisis.com.

Stiga, Unit 8, Bluewater Estate, Bell Close, Plympton, Plymouth PL7 4JH. Tel: 0845 600 3215. Website: www.stiga-online.com.

Stihl, Stihl House, Stanhope Road, Camberley, Surrey GU15 3YT. Tel: 01276 20202. Fax: 01276 670510. Email: postmaster@stihl.co.uk.

Suffolk, Gipping Way, Stowmarket, Suffolk IP14 1EY. Tel: 01449 742031. Fax: 01449 742214. Website: www.boschgarden.co.uk.

Suzuki, Steinbeck Crescent, Snelshall West, Milton Keynes, Buckinghamshire MK4 4AE. Tel: 0500 011 959. Fax: 0870 608 1305.

Toro, 1 Station Road, St Neots, Cambridgeshire PE19 1QH. Tel: 01480 226800. Fax: 01480 226801. Website: www.torouk.com.

Victa, Launton Road, Bicester, Oxon OX26 4UR. Tel: 01869 363636. Fax: 01869 363620. Website: www.barrus.co.uk.

Viking, Stanhope Road, Camberley, Surrey GU15 3YT. Tel: 01276 20202. Fax: 01276 670510. Website: www.vikingmowers.co.uk.

Villiers. Email: www.villiersinfo@isdm.co.uk.

Webb, Gipping Way, Stowmarket, Suffolk IP14 1EY. Tel: 01449 742031. Fax: 01449 742214. Website: www.boschgarden.co.uk.

Westwood, Great Haseley, Oxford. OX44 7PF. Tel: 01844 278800. Website: www.westwoodtractors.com.

Wolf Garden Ltd, Launton Road, Bicester, Oxfordshire OX26 4UR. Tel: 01869 363636. Fax: 01869 363620.

Yamaha, Sopwith Drive, Brooklands, Weybridge, Surrey KT13 0UZ. Tel: 01932 358000. Website: www.yamaha-motor.co.uk.

Lawnmower museum

British Lawnmower Museum, 106–114 Shakespeare Street, Southport, Lancashire PR8 5AJ. Tel: 01704 501336. Fax: 01704 500564. Website: www.lawnmowerworld.com.

This unique museum, situated in the classic Victorian seaside town of Southport, holds a collection of over 800 rare lawnmowers, a huge archive of grasscutting patents, blueprints, memorabilia, lawnmower manuals and handbooks, vintage spare parts and restoration workshops. The collection includes lawnmowers of the rich and famous, from Prince Charles and Princess Diana to Brian May and Hilda Ogden, as well as the most expensive and fastest lawnmowers in the world.

Lawnmower racing

British Lawnmower Racing Association, Hunt Cottage, Wisborough Green, Billingshurst, West Sussex RH14 0HN. Website: www.blmra.co.uk.

The North West Lawnmower Racing Association, 112 Western Drive, Leyland, Lancashire PR5 3JH. Website: nwlmra.co.uk.

US Lawnmower Racing Association. Website: www.letsmow.com/uslmra.